VISITING MACKINAC

VISITING MACKINAC

150 Years of Tourism at Michigan's Fabled Straits

Frank Boles

MICHIGAN STATE UNIVERSITY PRESS | *East Lansing*

Michigan State University Press
East Lansing, Michigan 48823-5245

Library of Congress Cataloging-in-Publication Data
Names: Boles, Frank, 1951– author
Title: Visiting Mackinac : 150 years of tourism at Michigan's fabled straits / Frank Boles.
Other titles: 150 years of tourism at Michigan's fabled straits
Description: East Lansing : Michigan State University Press, [2026] | Includes bibliographical references and index.
Identifiers: LCCN 2025017027 | ISBN 9781611865516 paper | ISBN 9781609177959 pdf | ISBN 9781628955583 epub
Subjects: LCSH: Tourism—Michigan—Mackinac Island (Island) | Tourism—Michigan—Mackinaw City | Tourism—Michigan—Saint Ignace | Tourism—Michigan—Mackinac, Straits of, Region | Mackinac, Straits of, Region (Mich.)—History | Grand Hotel (Mackinac Island, Mich.)—History
Classification: LCC G155.M654 B65 2026 | DDC 917.74/92304—dc23/eng/20250520
LC record available at https://lccn.loc.gov/2025017027

Cover design by Erin Kirk
Cover art is Arch Rock © Kevynbj | Dreamstime.com

Visit Michigan State University Press at *www.msupress.org*

For Nick

With appreciation and thanks for
the many journeys we have taken together.

Contents

PART 2. The Shoreline

Preface

The Straits of Mackinac have been a place that tourists have visited for more than 150 years. Over those many years, the location has developed traditions and lore that have been woven into a story. It is, however, a story told most often by the tourist industry itself, and as a result, has sometimes been turned into entertainment and at other times become self-serving. "Mackinac malarkey" is how individuals in the local community describe the most outrageous exaggerations that are passed off as fact. Every tourist area has its own version of Mackinac malarkey. But the malarkey offered at the Straits of Mackinac has had a very long time to develop and is particularly rich. It is has been honed and polished and, even when it is self-serving, can be very entertaining.

This volume attempts to sort through what has been written and said about tourism at the Straits of Mackinac and create a documented history for the tourist industry on Mackinac Island and in the communities of St. Ignace and Mackinaw City. In writing it, I have read much scholarly literature about the history of tourism, many books and articles, written by a wide group of authors, that seemed relevant to the straits tourist industry and researched the history of tourism at the straits in several newspapers. I have also had the opportunity to read many oral histories and unpublished reminisces found in libraries created both by people "in the know" and by people who shared their life stories. These are the sources used for scholarly research, but I hope the result you hold has been written in a way that will make it an enjoyable and informative read for the general public.

Although the Straits of Mackinac is a single geographic area, the tourist industry that developed there is not a single story. It is divided by geography. Mackinac Island has a longer tourist history than that found in the two cities that sit on the shoreline, St. Ignace in the Upper Peninsula and Mackinaw City in the Lower Peninsula. Because of this I have chosen to tell the story of tourism on Mackinac Island separately from the history of tourism on the shoreline. I have also focused on mass tourism at the expense of the more favored visitors to the island, people who arrive by private boat or airplane and the cottagers living in the large mansions along the island's bluffs.

It is important to add a reminder that this book is not a comprehensive community history of Mackinac Island, St. Ignace, or Mackinaw City. Because of this, many interesting stories are not included. For example, the island's snowmobile war of the 1970s is not discussed at length. Briefly, the war began when the Mackinac Island State Park Commission tried to apply the island's ban on motorized vehicles to snowmobiles. If it had a motor, reasoned the commission, it was banned from park land. Year-round residents, offended that the "summer people" did not understand the hardships of life on the island during the winter, demanded the commission reverse itself. The revolt was highlighted by an act of civil disobedience. Snowmobilers gathered just inside the city limits (where their machines were legal) and as a group violated commission rules by racing into the park, ultimately driving their snowmobiles in circles around a besieged Fort Mackinac. It is a good, and true, story, but it is not about tourism and thus not a story told in this book. (But since I have mentioned the war, I should tell you how it ended: The commission ultimately surrendered. During the winter the island's permanent residents drive their snowmobiles on park land.)

As I have already written, I have employed historical research and academic literature in creating this volume, and while I hope it might prove of use among academics who study tourism, the book was not written for academics. It was written for the general public. I hope you enjoy it.

Acknowledgments

In the writing of this book I have found myself in debt to a great number of people who facilitated my research.

Leroy Barnett, the dean of Michigan's bibliographers, was a staunch ally. Over many years as director of the Central Michigan University's Clarke Library, I learned that when we simply could not find an answer regarding something, it was time to invoke the phrase "Talk to Lee." I did indeed talk to Lee at a critical juncture in this project, and found him the master of obscure sources hiding important information. Lee is a treasure.

Michael Federspiel, an old friend, proved a valuable sounding board as the project progressed. His own published works about the Petoskey area gave him particular insights into the writing of this book. I thank him for his help.

The staff of the Michigan State Archives in Lansing, Michigan, were helpful in my research of public records. Their aid in finding and rapid retrieval of material was of great help in creating this book. I offer them my sincere thanks.

The staff of the Bentley Library graciously assisted me in my research, in particular by "bending" rules regarding the number of boxes of material that could be retrieved during the course of the day. I reviewed the Mackinac Island State Park Commission files found in the papers of Governor G. Mennen Williams. The Williams collection, however, is arranged chronologically and within the annual files, by alphabet. The commission's records were found in a large number of boxes. Because I was interested in what was usually one, relatively thin, folder in each box I needed, it sometimes took only a few minutes to examine the file. By allowing me to view more than the routine allotment of boxes I was able to completely go through the large number of boxes I needed to review in an afternoon, rather than the three days it likely would have otherwise taken. I offer my sincere thanks to the Bentley Library staff for this help.

Local resources in the Straits of Mackinac region proved of great importance. The Cheboygan Area Public Library held resources in their local history collection not available in other nearby libraries. My thanks to the library for maintaining this valuable resource. The staff of the St. Ignace Public Library was helpful, but the lack of a significant literature about the history of that community limited the information they could offer to me. Fortunately, Erin Vanier of the Michilimackinac Historical Society in St. Ignace was able to provide images for this volume, for which I am grateful. The Michigan History Room of the Mackinaw Area Public Library, located in Mackinaw City, was a rich resource of information. I am very much in debt to the library's staff, and in particular the library's director, Tanya Procknow.

Brian Jaeschke of the Mackinac Island State Park Commission was gracious and helpful in facilitating my use of the Keith R. Widder Library, which is housed in the commission's office building located in Mackinaw City. Open to the public only by appointment, Brian made sure the door was unlocked when I needed to peruse the commission's records and offered valuable advice about resources in the library that I had not been aware of. For all this, I thank him very much.

For the many newspapers used in researching this work, I am indebted to both the Library of Michigan and the Clarke Historical Library. The Library of Michigan has, over the course of generations, collected on microfilm a vast number of Michigan newspaper titles. Many of these titles have been digitized and made available online by the Clarke Historical Library through the Clarke Library's own website, and many of the Michigan newspapers have been made available online through the Library of Congress. Other newspaper titles have never been digitized and remain available only on microfilm. The Library of Michigan offers a uniquely valuable service by gathering microfilmed newspapers from throughout Michigan in Lansing, where the public can view them.

I must offer thanks also to many of my former colleagues at the Central Michigan University Libraries. Let me begin by thanking Dean of Libraries Kathy Irwin for making available to me, after my retirement, a faculty study room within the Park Library Building. The space was a valuable one in which to read, reflect, and write.

I am also in debt to the Library's Documents on Demand unit for finding publications that have not been digitized or were digitized in a format that proved hard to use. For those who believe that everything is on the internet, or at least everything worth reading, I offer testimony that this is not so. Important information is found in publications no one has yet digitized or microfilmed. It is available only through extensive travel or by the work of those who maintain the nation's interlibrary loan structure. My thanks to CMU's Documents on Demand team, who maintain CMU's inter-library loan operation.

I also owe a very deep debt to the Clarke Historical Library and its excellent reference librarian, Bryan Whitledge. The Clarke's rich printed collection was surprisingly complete regarding works published about the straits. Beyond looking at the Clarke Library's books, I spent months in the library's reading room, slowly working my way through microfilmed newspapers that had never been digitized. Bryan Whitledge always went over and above what a researcher could hope for in making the library's resources available to me and attempting to answer my sometimes obscure questions. I am deeply in debt to Bryan, who is among the best reference librarians I know.

And thank you to my wife, Valerie. She has put up with years of annoying conversations about trivial details of straits history, "deferred" household tasks that were, in fact, never completed, and the more than occasional postponement of something she would much rather have been doing. And if that were not enough, she read the whole book in draft and made both large and small suggestions for improvements, such as thoughtful ideas for revision and finding an amazing number of missed periods. It was a thoughtful, and thorough, review. She is a saint.

All these institutions and individuals have my lasting thanks, and although it probably need not be said, let me say that none of them bear any responsibility for errors of fact or interpretation that exist in the following pages. The mistakes are mine, not theirs.

PART 1

The Island

CHAPTER 1

The Pioneer Tourists, 1836–1861

The first visitors to the Straits of Mackinac who came simply to see Mackinac Island arrived in the 1830s. Many people who traced their ancestry to Europe had come before them, but they had come either to do business on the island or were sent there by various governments to ensure the orderly conduct of business and to establish a military presence. The island had been prominent first in the fur trade. When the Great Lakes fur trade came to its end after the War of 1812, it was replaced on Mackinac Island by a successful commercial fishing industry.[1]

While most of the island's residents worked in this mix of a declining fur industry and an emerging commercial fishing center, the first tourists appeared. These pioneer tourists, many of whom were writers, stepped ashore in the 1830s. The first influential writer, who came for no other reason than to see what was to be seen, was Harriet Martineau. Martineau was a prolific English author. She had published well-received volumes of social commentary about Britain, and in 1834, Martineau sailed to America, where she planned to write a similar book about American society.[2] While in southern Michigan, she had been encouraged to visit

Mackinac Island.[3] On July 5 the ship in which she was traveling stopped at the island to deliver a small quantity of cargo. The captain planned to quickly unload the cargo and resume his voyage. But the ship arrived near dusk and the person responsible for receiving the cargo insisted the shipment stay onboard, until it could be more safely unloaded in the morning. Grudgingly, the captain agreed to spend the night. Taking advantage of the situation, Martineau made arrangements to go ashore at dawn.

Martineau was up at 5:00 a.m. for the trip. The commander of Fort Mackinac gave her a brief tour of the island, took her for breakfast with his family, and had her back aboard the ship when it sailed at about 9:00 a.m. Martineau spent at most four hours on the island.[4] It was, however, an extraordinarily influential four hours. Her account of the visit, published in 1837 in her widely read *Society in America*, portrayed Mackinac Island as a place of mythical beauty.

"The Island looked enchanting as we approached, as I think it always must, though we had the advantage of seeing it first steeped in the most golden sunshine that ever hallowed lake or shore."[5] Mackinac was a place of "melting beauty which one almost fears will vanish as its softens before one's eyes"; Martineau found the harbor enchanting: "Nothing could be more bright and joyous." "I can compare it [Mackinac Island] to nothing but to what Noah might have seen, the first bright morning after the deluge."[6] If the island's beautiful scenery was comparable to what Noah saw, the healthful environment was comparable to the one Noah himself enjoyed, a man who the Bible says lived 950 years.[7] Martineau wrote, likely with a smile and a wink, "The Island is so healthy that, according to the commandant, people who want to die must go somewhere else."[8]

Architecture was another topic Martineau addressed, but unlike the island's natural beauty, she had nothing nice to say about the town that had been built on it. "The houses of the old French village are shabby-looking, dusky, and roofed with bark." If the town disappointed her, the souvenirs did not. She took away "some pretty purchases of Indian manufactures . . . light matting of various colours, and small baskets of birch-bark, embroidered with porcupine quills, and filled with maple sugar."[9] Sweet treats and a few pretty trinkets to show the folks back home assumed their place as part of the island's tourist experience.

When sailing away from Mackinac Island, Martineau summarized her experience. "We were in great delight at having seen Mackinac, at having the possession of its singular imagery for life; but this delight was at present dashed with the sorrow of leaving it. I could not have believed how deeply it is possible to regret a place, after so brief an acquaintance with it."[10]

A long list of authors followed Martineau to Mackinac Island. Like Martineau, many of them funded their travel by publishing books about their journey. Travel narratives were a well-established literary genre. Between 1800 and 1850 about 325 travel guides were published in the United States. Readers of the day enjoyed these books. For example, in the late 1840s, travel books amounted to 19 percent of the books checked out by men and 15 percent of the books checked out by women from the New York Society Library.[11] At least seventeen authors who visited Mackinac Island before the Civil War published their observations about the place,[12] but four writers were particularly important. The published works of Harriet Martineau (1837), Anna Brownell Jameson (1838), Frederick Marryat (1839), and Margaret Fuller (1843) first defined Mackinac to the reading public.

Harriet Martineau was the first to arrive, but not far behind her was Anna Brownell Jameson. She had published her first, very successful travel diary in 1826, beginning a long career as an author.[13] Jameson came from England to Canada in 1836. Her husband was a government official in Toronto, but the marriage was unhappy, and Anna preferred to remain in England. However, her husband hoped for a more senior Canadian government post. To get it, social convention required his "lady" be present. Therefore, he demanded Jameson join him in Canada. She arrived in Toronto in November 1836. She left for England in September 1837. Her baggage included the makings of a book, along with papers legally separating her from her husband.[14]

For practical purposes, the divorce likely began in June 1837, when Jameson left her husband's house in Toronto and traveled through the Great Lakes. The journey, along with writing done while in Toronto, served as the basis of a three-volume work, *Winter Studies and Summer Rambles*, published in London in 1838 to critical acclaim and popular success.[15] Like Martineau, whom she had read,[16] Jameson visited Mackinac Island.

Like Martineau, Jameson began with words of praise regarding the island's appearance. Arriving at night, she awoke and realized the ship had anchored at Mackinac Island.

> I dressed in all haste and ran up to the deck, and there a scene burst at once on my enchanted gaze, such as a I never had imagined, such as I wish I could place before you in words—but I despair, unless words were of light, and lustrous hues, and breathing music. . . . I cannot tell how long I may have stood, lost—absolutely lost, and fearing even to wink my eyes lest the spell should dissolve, and all should vanish away like some air-wrought phantasy some dream out of fairy land.[17]

Jameson was particularly interested in the Native American population. Jameson's opportunity to observe the Indigenous population was substantially enhanced because she was a guest of Indian Agent Henry Schoolcraft and his wife, Jane Johnston, during the period when he was distributing treaty required annuities.[18] A constant stream of Native visitors came to Schoolcraft's door to receive the money the US government was obliged to give them.

Her descriptions of Native dress and appearance are long and foretold her later work as an art critic. Of one young man, she writes:

> His face was most artistically painted, the upper half of it being vermilion, with a black circle round one eye, and a white circle round the other; the lower half of a bright green, except the tip of his nose, which was also vermilion. His leggings of scarlet cloth were embroidered down the sides, and decorated with tufts of hair. The band, or garter, which confines the leggings, is always an especial bit of finery; and his were gorgeous, all embroidered with gay beads, and strings of tassels of the liveliest colours hanging down to his ankle. His moccasins were also beautifully worked with porcupine quills; he had armlets and bracelets of silver, and round his head a silver band stuck with tufts of moosehair, dyed bule and red; and, conspicuous above all, the eagle feather in his hair, showing he was a warrior, and had taken a scalp—i.e. killed a man. Over his shoulders hung a blanket of scarlet cloth, very long and ample, which he had thrown back a little, so as to display his chest, on which

> a large outspread hand was painted in white. It is impossible to describe the air of perfect self-complacency with which this youth strutted about.[19]

If the appearance of the Indigenous population was something Jameson wrote about in positive terms, Jameson was less insightful about Native American culture, employing the common nineteenth century viewpoint that Native Americans were a noble but fallen people. Jameson wrote that she heard "shouts and shrieks of the intoxicated savages, who were drinking in front of the village whiskey store," and she saw

> one of the Ottawa Chiefs, . . . his name is Kim,E,Wun, which signifies the Rain, . . . he now stood before me, one of the noblest figures I ever beheld. . . . He stood in a contemplative attituded evidently undecide whether he should join his drunken companions in their night revel, or return, like a wise man, to his lodge and his mat. He advanced a few steps, then turned, then paused and listened—then turned back again. . . . Alas it was soon decided—the fatal temptation prevailed over better thoughts. He suddenly drew his blanket round him, and strided onwards in the direction of the village, treading the earth with an air of defiance, and a step which would have become a prince.[20]

Following Martineau and Jameson was Captain Frederick Marryat, a British author whose books in the 1830s were among the most widely read in both Britain and the United States. Marryat, who was on his own American tour, arrived at Mackinac Island on August 1, 1837. He left on August 9, but his stay was interrupted by a canoe trip to Sault Ste. Marie.[21] In a book published in 1839, *Diary in America*, he briefly describes his visit to Mackinac Island.

Marryat understood he had little new to say and thus passed over Mackinac Island quickly in his book. But in his few words reinforced the island's positive image. "Miss Martineau has not been too lavish in her praises of Mackinaw [*sic*]." And like Martineau, Marryat heard and repeated the tall tale about how there were no sick people there. In his telling of the story, Marryat began by noting the portly condition of the Americans on the island. When he asked about it, he was told, "We have good air, good water, and what we eat agrees with us.'" This was followed by the inevitable joke about death: "If people want to die, they can't die here—they're obliged to go elsewhere."[22]

If Marryat was largely content to refer his readers to Martineau's observations and repeat the already well-worn joke about the island's healthfulness, he added to Martineau's description a metaphor that was first alluded to by Jameson but put into a phrase by Marryat that generations of subsequent writers would borrow, with or without attribution. "It [Mackinac Island] has the appearance of a fairy isle floating on the water, which is so pure and transparent that you may see down to almost any depth; and the air above is as pure as the water, so that you feel invigorated as you breathe it."[23] Descriptions penned by nineteenth and early twentieth century authors included references to fairies, or simply borrowed the phrase "fairy isle," with great regularity. In truth, what Marryat said about Mackinac Island was both brief and largely repetitive of Martineau, but that it was said at all by one of the most read authors of the day made it very influential.

In 1843, Margaret Fuller visited the island for nine days as part of a summer she spent traveling across the Great Lakes. Although Fuller did not have the wide popular audience of Martineau, Jameson, or Marryat, she was well connected in New England literary circles. She had translated much German romantic literature into English and had served as editor of the influential transcendentalist magazine, *The Dial*. Her audience was not wide but it was significant. Thus, her publication, *Summer on the Lakes* (1843), was read by people whose opinions mattered.[24]

Because she was writing for a well-read audience, Fuller knew they likely already had read descriptions of the island's natural beauty. Thus, she went out of her way to lower expectations regarding her own description of the scenery, saying that "Mackinac has been fully described by able pens, and I can only add my tribute to the exceeding beauty of the spot and its position." Despite this statement, Fuller wrote a rather long description of the island that largely repeated and reinforced what Martineau had written. Fuller distinguished herself, however, when she turned to the subject of the Native population. An early feminist, who closely observed social customs within New England, Fuller was equally interested in social customs within Native communities. Fuller's writing about Native Americans is the most nuanced of the pre–Civil War authors who visited

Mackinac Island. She rejected stereotypes, writing, "Why will people look only on one side? They either exalt the Red man into a Demigod or degrade him into a beast."[25] She understood that what she observed reflected the largely negative consequences on Native culture that resulted from contact with Europeans and their American descendants.

> I feel that I have learnt a great deal of the Indians, from observing them even in this broken and degraded condition. There is a language of eye and motion which cannot be put into words, and which teaches what words never can. I feel acquainted with the soul of this race; I read its nobler thoughts in their defaced figures. There was a greatness, unique and precious, which he who does not feel will never duly appreciate the majesty of nature in this American continent.[26]

Fuller's first impression of the Native Americans is one of industry and artistry. "They were already on the alert [at dawn's first light], the children creeping out from beneath the blanket door of the lodge; the women pounding corn in their rude mortars, the young men playing on their pipes." Of the pipe playing she notes that when she had heard the same tunes played on a Western instrument she found it difficult to imagine "any one fancying it a melody; but now, when I heard the notes in their true tone and time, I thought it not unworthy comparison, in its graceful sequence, and the light flourish, at the close, with the sweetest bird-songs."[27]

Her second impression, taken in the late afternoon, is similar, although with more than a hint of the theme of a fallen people.

> The Indians were grouped and scattered among the lodges; the women preparing food, in the kettle or frying pan, over the many small fires; the children, half-naked wild as little goblins, were playing both in and out of the water. Here and there lounged a young girl, with a baby at her back, whose bright eyes glances, as if born into a world of courage and of joy, instead of ignominious servitude and slow decay. Some girls were cutting wood, a little way from me, talking and laughing, in the low musical tone, so charming in the Indian women.... It was a scene of ideal loveliness" one to which she was "constantly attracted.[28]

Particularly interested in family relationships, she writes a positive portrayal of Native Americans and their families. Of the women she writes,

> Their decorum and delicacy are striking, and show that when these are native to the mind, no habits of life make any difference. Their whole gesture is timid, yet self-possessed. They used to crowd round me, to inspect little things I had to show them, but never press near; on the contrary, would reprove and keep off the children. Anything they took from my hand was held with care, then shut or folded, and returned with an air of lady-like precision. They would not stare, however curious they might be, but cast side-long glances.[29]

Surprisingly, since she generally argued that a woman's lot in life was harder than that of a man, Fuller goes to some length to disagree with the popular notion that a Native man "compels his wife to do all the drudgery, while he does nothing but hunt and amuse himself." To make her case, Fuller frames the hunt as central to the family's existence. With this as her foundation, she writes,

> When he [the man] returns from hunting, worn out with fatigue, having tasted nothing since dawn, his wife, if she is a good wife, will take off his moccasins and replace them with dry ones, and will prepare his game for their repast, while his children will climb upon him, and he will caress them with all the tenderness of a woman; and in the evening the Indian wigwam is the scenes of the purest domestic pleasures. The father will relate for the amusement of the wife, and for the instruction of the children, all the events of the day's hunt, while they will treasure every word that falls, and thus learn the theory of the art, whose practice is to be the occupation of their lives.[30]

That Fuller and Jameson published so many words regarding the island's Indigenous population demonstrated the interest people had in the life of America's Native people. Many readers would buy their books to address that interest, but a few would go a step beyond and come to Mackinac

Right: Margaret Fuller, *Summer on the Lakes* (Boston, 1844), frontispiece of illustrated edition. Hathi Trust, courtesy Clarke Historical Library, Central Michigan University.

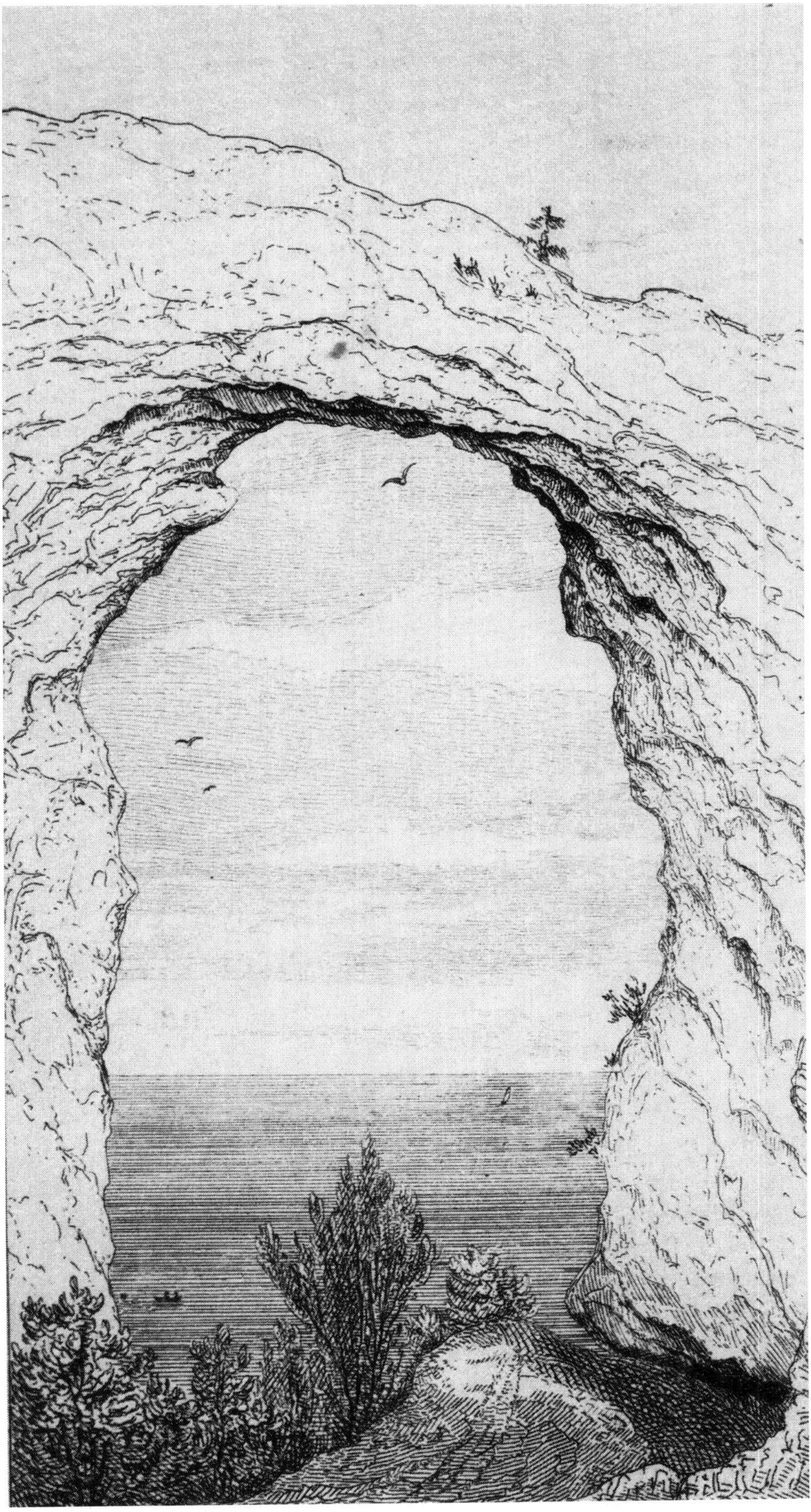

Island, hoping to encounter Native Americans or at least obtain from the local merchants an item made by the Anishinaabe people.

Writing fascinating tales about the Native American population, along with lavish descriptions of the island's natural beauty and the healthful environment were part and parcel of the portraits painted in words by the pioneer writers who visited the island. Mackinac Island's pioneer illustrators also cast a beautiful glow around Mackinac. Long before the first photograph was taken on the island in 1856, many artists had visited and some of their work had been published. The first widely accessible image appeared in Charles McKenney's 1827 book, *Sketches of a Tour of the Lakes.*[31] McKenney was a government official, sent to the area to deal with Native American concerns. He took along an artist, J. O. Lewis, who likely created the image portraying the island from the water. The image was drawn from a considerable distance and thus did not include in any detail about human-built features, other than the fort. Among the early and influential travel books describing Mackinac in words, Margaret Fuller's publication, *Summer on the Lakes*, was unique in that it also included illustrations of the island. Fuller was so impressed by the island's scenery that she chose a view of Arch Rock, one of the most striking geological features of the island, as the frontispiece for the illustrated version of her book, drawn by her traveling companion Sara Ann Clarke. Fuller also included in her book two other island images drawn by Clarke, a second image of Arch Rock, and a beach scene portraying Native Americans, with teepees and canoes.[32]

During the 1850s, illustrations and stories about the island began reaching a new and larger audience. Emerging weekly newspapers developed a national readership within the United States. These publications often featured what today would be called dream vacations. Mackinac Island found its way into more than one dream. *Ballou's Pictorial Drawing-Room Companion* printed a well-illustrated feature about the island on May 5, 1855. The article included three illustrations: the town as viewed from the water, Arch Rock, and Sugar Loaf Rock. *Frank Leslie's Illustrated Newspaper* ran an article about Mackinac Island on September 20, 1856, that also included three illustrations.[33]

By the 1850s, the steamships frequently passing through the Straits of Mackinac brought people who had read about Mackinac Island to see the island's beauty. The first steamship reached Mackinac Island in 1819.[34] By 1824, four steamboats a week were arriving. A quarter century later, in 1847, as many as four steamboats a day might arrive at the island.[35] Ships stopped both for fuel and some fresh fish to be served at supper, and usually offered passengers a short shore excursion during the time needed to load the wood to burn and fish to eat.[36] A few of the passengers who had read about the island and seen illustrations of it stayed for more than an hour or two. This was so because one of the nation's most influential tourist guidebooks, published by D. Appleton of New York City, had begun to take note of the island. Appleton's earliest national guidebook, published in 1846, mentioned Mackinac Island, but from the perspective of a voyager seated on the deck of a Great Lakes passenger ship. It never mentioned what the traveler might see if they got off the boat.[37] The company's 1850 guide also gave a brief description of the village on the island, but again mostly to inform a ship's passenger about the community coming into view.[38] The 1851 guide, however, recommended that a person come ashore, if only for a day.

> Saturday morning finds you moored amidst bark canoes and Indian tents, close under a high rocky fort, bristling with a hundred cannon. After getting a breakfast of Mackinaw trout, trying your luck at fishing in water so clear that you can see a trout twenty feet from the surface, buying a few trinkets from the young squaws, visiting the forts and barracks.[39]

A day visitor might also enjoy a carriage ride. By the 1840s, a fairly consistent carriage tour route existed that incorporated natural and historic sites.[40] The 1857 edition of the Appleton guidebook was the first edition that recommended spending the night on the island. Appleton's 1857 guidebook informed the reader that the island was "picturesque and romantic in the highest degree." However, Mackinac Island remained the preserve of the elite tourist, as the recommended two-day stay was part of a larger four-week tour of the Great Lakes.[41]

The descriptions found in the Appleton guidebooks reflect the beginnings of a true tourist industry on the island. Guests were arriving and in need of accommodations. Boarding houses had begun to appear in the late 1830s.[42] In the 1840s, the Franks family was hosting guests in the former Protestant Mission House, which had the reputation of being the island's finest overnight accommodation.[43] In 1852, Charles O'Malley opened the Island House Hotel, and in 1858, Reuben Chapman opened the Lake View House.[44]

By the late 1850s, Mackinac Island could boast of several hotels, but it could not boast that any of them were particularly good. All the hotel owners had built their lodging as much to accommodate men going to or coming from the newly opened mining operations in the Upper Peninsula as they did for tourists. The hotels were relatively small and often overcrowded. Margaret Fuller arrived expecting "a room to myself at the hotel, but found none, and was obliged to take up my rest in the common parlor and eating-room, a circumstance which insured my being an early riser."[45] William Cullen Bryant, who in 1846 stayed in Mission House, described it as "a plain, comfortable old wooden house, built thirty or forty years since, by a missionary society, and now turned into an hotel."[46] If comfortable, Mission House was also crowded. "Parlors and garrets were turned into sleeping-rooms; beds were made on the floors and in the passages, and double-bedded rooms were made to receive four beds."[47] In 1850, the famed naturalist Louis Agassiz led an expedition from Boston to Lake Superior, which made a brief stop at Mackinac Island. Leaving their steamer, the party found a single street of "struggling log cabins and ill conditioned frame houses, parallel with the beach, and some of better class standing back among gardens at the foot of the bluff." As for hotels, the Bostonians only deemed Mission House acceptable, although there is the impression that they needed to lower their East Coast standards to spend the night even there.[48]

East coast residents may have been unimpressed by Mackinac Island's village, but local entrepreneurs sold them souvenirs, just as they had to Harriet Martineau. At the heart of much of the pre–Civil war souvenir trade was real or faux Native American artifacts. Daniel Curtiss, in 1852, noted "Indian fabrics" as something "purchased by visitors and passengers, while the boats make their short stops."[49] They were readily available and were supplemented by various baked goods.[50]

The incorporation of Indigenous products into tourism usually was done in a way that did not honor the culture and people who had created it. A description written by a member of a group of immigrants headed toward Wisconsin notes that on their brief visit to Mackinac Island among the first signs they saw was one painted in large letters reading "Injun Curiosities." It was still there in 1848 when another visitor commented about it.[51] And if "Injun Curiosities" failed to entice money out of visitors' pocket, the merchants also offered for sale the same sweet treats Martineau had purchased, small containers filled with maple sugar called moluks.[52]

By the late 1840s and into the 1850s, enough people had made a visit to Mackinac Island that Charles Lanman wrote in *A Summer in the Wilderness* (1847), "I now write from Mackinac, the beautiful, which studs the waters of the north, as does the northern star its own cerulean home. But what can I say about this Island that will be new, since 'every body' now pays it a brief visit while journeying in the West."[53] Other authors also began to worry about how tourism might impact the island's future.

In 1838, Jameson was perhaps the first to express concern about the impact of tourism. "The facility of communication lately afforded by the lake steamers, and its situation half way between Detroit and the newly settled regions of the west, are likely to render Mackinac a sort of watering-place for the Michigan and Wisconsin fashionables, . . . How far such an accession of fashion and reputation may be desirable, I know not; I am only glad it has not yet taken place, and that I have beheld this lovely Island in all its wild beauty."[54] In 1846 another author, William Cullen Bryant, wrote in a similar vein, "I can not [*sic*] think with a kind of regret on the time which, I suppose is near at hand, when its [Mackinac Island's] wild and lonely woods will be intersected with highways, and filled with cottages and boarding houses."[55]

The concerns of Lanman, Jameson, and Bryant, however, were far ahead of reality. Mackinac was a distant destination drawing only a small group of tourists and almost accidental visitors who happened to be on a passing steamer that stopped at the island's harbor. Most pre–Civil War tourists who came to see Mackinac Island were wealthy, but by the 1820s, most wealthy tourists followed a well-defined "grand tour" of America that came nowhere near Mackinac Island. Their travel centered around New England and New York, ending its westward leg at Niagara Falls, which was America's premier tourist destination throughout the nineteenth century.[56] "Every body" was not coming to Mackinac Island. Steamships, however, had made travel on the Great Lakes sufficiently easy that an adventurous few travelers did come to Mackinac. As a result, the remote island that had been described lovingly by pioneering travel writers had, by 1860, begun to evolve into something other than what they had seen. But the transformation was slow, and the island would not become a true resort until well into the late nineteenth century.

CHAPTER 2

Becoming a Fashionable Resort, 1870–1894

The opening of the Grand Hotel, on July 10, 1887, signaled Mackinac Island's arrival as one of the Midwest's leading resorts.[1] Becoming a premier tourist destination was a difficult journey. In the 1870s, the island's commercial fishing industry was dying. In 1872, Mackinac Island packed and shipped only a few thousand barrels of fish, compared with the quarter million barrels of fish that had left the island each year before the Civil War. If the island economy was to survive, a new industry needed to be found.[2] The only choice was tourism.

Tourism had begun before the Civil War, but little had been done to prepare the island for a future based solely on tourism. What Mackinac Island offered tourists in the 1870s was largely what had been available in the 1850s: a healthy environment, the island's natural beauty, and a small village with meager accommodations for visitors. Despite the state of the village, health and beauty drew many visitors. The idea of traveling to improve health was well established.[3] After the Civil War, the number of people traveling to improve their health increased dramatically because of two "new" diseases, neurasthenia and hay fever.[4]

As the nineteenth century progressed, urban dwellers, who in all other ways appeared healthy, nevertheless complained to their doctor of pain, emotional distress, and exhaustion. To explain these symptoms, from 1870 to about 1915 many physicians diagnosed patients with neurasthenia. It supposedly was caused by the unique stress created by living in an urban, industrial environment. Diagnosis of the disease reached near epidemic proportions among white middle- and upper-class men by the 1880s.[5]

Hay fever, as we understand it today, was first described in 1819. By the 1860s, hay fever was also frequently diagnosed. While anyone might suffer from hay fever, doctors believed people of high status and above average income were particularly inclined to suffer from it. Whether a person was diagnosed with hay fever or neurasthenia the cure was the same: a trip to the mountains or the seaside.[6]

With its reputation for healthy living, Mackinac Island benefited from travelers diagnosed with either neurasthenia or hay fever, but the island had no exclusive claim to cure these aliments. Virtually every American tourist destination made some claim about improving health. Even places that today would not appear of much benefit to the sick claimed otherwise. Proponents of Mammoth Cave in Kentucky asserted that people with consumption (tuberculosis) could be cured by renting a cabin constructed inside the cave itself. The ill individual would supposedly benefit from the restorative cavern air. The cure did not attract many sufferers, likely because the sick individual would also live in complete darkness, except for the glow of an oil-fired lantern. The mineral springs at Yellowstone National Park were promoted as helpful to those with rheumatism or skin diseases. The foul smell of sulfur and sulphurated hydrogen that permeated much of the area where the springs were located supposedly proved the location's restorative properties.[7] In the nineteenth century, Mackinac Island's claim to bestow good health was in competition with many other tourist destinations in northern Michigan, the upper Great Lakes, and much of America.

Mackinac Island was beautiful, but so were many other destinations from which a tourist could choose. And many of those other destinations had accommodations undreamed of on Mackinac Island. For example, in the summer of 1873, Louisa Roberts and her husband traveled from

Village of Mackinac Island, 1875. John Disturnell, comp., *Island of Mackinac, Giving a Description of All the Objects of Interest and Places of Resort* (Philadelphia, 1875). COURTESY CLARKE HISTORICAL LIBRARY, CENTRAL MICHIGAN UNIVERSITY.

Philadelphia to Lake Erie's Put-In-Bay Islands. The location featured several hotels, one of which could accommodate over a thousand guests. The Roberts were surprised to find that the hotels were full to overflowing with "thousands" of visitors boating, swimming, and fishing.[8] Mackinac Island's tourist infrastructure lagged far behind its Lake Erie competitor.

Further complicating the island's images was that for the first time, travel narratives were saying unkind things about the island's village. Some books published in the 1870s followed the tradition of printing nothing but positive statements about Mackinac Island. Appleton Company's 1871 guidebook stated that the island offered a "pretty little village at the foot of the cliff" that featured "several good hotels and several stores where Indian curiosities, petrifications, agates, photographs and other mementos" could be purchased. But other books were devastating in their criticism of the village.[9]

Picturesque America, published by the Appleton Company in 1872, included a scathing description of the village.

> The village at the foot of the cliff, is decayed and antiquated; the fort, on the height above, is white and crumbling with age; the very flag is tattered; and once, beyond the fringe of habitation around the port, there is no trace of the white man on the Island save one farmhouse of the last century; and a ruin on the western shore. There is no commercial activity at Mackinac; the business life of the village died out with the fur trade; and so different is its aspect from the other lake-towns, no matter how small, that the traveller feels as though he is walking through the streets of a New-World Pompeii.[10]

Whether the island's village was pretty or a "New-World Pompeii" likely depended upon the eye of the traveler, but what was accurate to say was that in the early 1880s, much of the village was unchanged from the 1820s. Most visitors believed the best accommodations on the island remained at Mission House, although some preferred Island House. Mission House, however, had not changed that much from what had existed in the 1840s, with bedrooms little bigger than closets. As late as 1884, one widely distributed tourist volume wrote that the island's hotels offered accommodations of little distinction. An overnight guest could expect "a one-course dinner, purely outdoor entertainment, Spartanly furnished rooms, and little in the way of service." The accommodations on Mackinac Island offered few, if any, luxuries.[11]

If a visitor nevertheless did choose to spend the night on Mackinac Island, what they likely did was take a carriage tour of the island to experience its natural beauty and then explore the town's twelve to fifteen stores to purchase sweets or souvenirs.[12] The stores used old and new ideas to persuade tourists to part with their money. Just as they had in the first half of the nineteenth century, the lives and crafts of the region's Indigenous people continued to fascinate those who came to visit the area. And much like in the days before the Civil War, island stores catered to this interest.

S. Highstone advertised a "New Indian Curiosity Store" which featured "the largest, best assorted and cheapest stock of Indian Curiosities ever opened on the Island of Mackinac." There, a visitor could buy bark boxes, trunks, baskets, sweetgrass, beadwork, feathers and "every description of Fancy Work made by Chippewa Indians." Highstone, however, had a competitor, named the "Mackinac Indian Bazaar." The Bazaar invited "Strangers & Visitors To this charming and Romantic Island, for ages the Paradise of the Genii, so believed from time immemorial by the Indian Tribes, wishing to procure a Souvenir of their visit to the 'FAIRY ISLE,' the spiritual resting-place of the departed Braves and renowned Warriors of the West, will fine a rare and splendid collection of GENUINE INDIAN CURIOSITIES." Although the phrasing of the ad is jarring to a modern reader, the Indian Bazaar made clear, in all capital letters, that it had "GENUINE INDIAN CURIOSITIES," superior, one must conclude, to the Chippewa "fancy work" found at Highstone's nearby shop.[13]

Selling Native American souvenirs was an old practice, but island merchants also offered the latest in modern technology. Photo souvenirs, often pictures of family groups visiting the island, were being sold at Mackinac in the 1860s.[14] In the late 1880s, locally made, European style candy began to be sold, competing with the maple sugar treats that had long been offered to visitors. Among the European treats was something new: fudge.[15] Fudge was not invented on Mackinac Island. It had been sold in New England in the mid-nineteenth century.[16] The first person to make fudge on Mackinac Island was Rome Murdick, the son of Henry Murdick. Henry was an experienced candy maker. He had begun making candy in 1877 in Mt. Clemens, Michigan. He moved his store several times before he came to Mackinac Island in 1889 and opened "Murdick's Candy Kitchen."[17]

Henry was also a ship builder, and after he arrived at Mackinac Island, he also opened a boat livery. Very quickly he decided that he would run the boat business and his son Rome would handle the candy store. Rome perfected and first sold Mackinac Island fudge. Despite the status it would eventually obtain, in the nineteenth century, fudge, as well as the whole idea of candy made on the island, enjoyed only modest success. Rome Murdick had the island "made on site" candy business to himself for a decade before a competitor, "Davis & Sons, Old Stand," opened in 1899.[18]

If Mackinac Island's tourist industry in the 1870s and 1880s did not appear destined for singular success, various national trends suggested it would, nevertheless, grow. In the 1870s, the number of people taking vacations increased dramatically.[19] Before the Civil War, a vacation was something only the rich could afford. But after the war, the new middle

Village of Mackinac Island between 1880 and 1899. Detroit Publishing Co. COURTESY THE LIBRARY OF CONGRESS.

class had enough excess income to indulge themselves in previously unenjoyed luxuries. Taking a vacation became a favorite middle-class activity. Going on vacation was made more possible not only because people had more money but because of the growth of railroads. In the twenty-five years after the Civil War, the amount of railroad track in the United States expanded fourfold.[20]

In 1881, the Michigan Central Railroad was the first to reach the Straits of Mackinac, with track coming from Cheboygan. Worried that the Michigan Central might corner the business at the straits, the Grand Rapids & Indiana Railroad (GR&I) reached Mackinaw City from the west in 1882. In the Upper Peninsula, the Detroit, Mackinac & Marquette Railroad (after 1887, renamed the Duluth South Shore & Atlantic Railroad, and usually called the Soo line) was the first railroad to reach St. Ignace.[21]

Even before railroad tracks reached Mackinaw City, the GR&I was selling vacationers tickets to the straits. Beginning in 1875, the railroad operated seasonal overnight sleeper trains six days a week from Cincinnati to points north. The "Mackinaw Express" left Cincinnati at 7:00 a.m. and ended its run when it arrived in Petoskey at 8:00 a.m. the next morning. The "Petoskey Express" left Cincinnati for Petoskey at 7:00 p.m. When either train arrived, a boat was waiting to complete the trip to Mackinac Island. In the 1870s, the Michigan Central Railroad did not have access to a lake port in northern Michigan. Nevertheless, it considered travel to the island sufficiently important to offer tourists a sleeping car service to Mackinac Island. To do this, the Michigan Central's Mackinac Island bound car traveled west from Detroit through southern Michigan to Grand Rapids. There it was detached from the Michigan Central train and added to the GR&I's Mackinaw Express.[22]

With tourist trains running almost every day, the railroads tried to sell tickets by giving away tourist publications, many of which included generous descriptions of Mackinac Island.[23] The Grand Rapids & Indiana Railroad was a leader in printing tourist books. Between 1875 and 1897, the railroad printed at least twenty-one substantial booklets advertising tourist destinations along its route. It would continue doing so, although with less frequency, until 1917.[24]

The GR&I's 1879 offering, *A Guide to the Health, Pleasure and Fishing Resorts of Northern Michigan*, gives a glimpse of Mackinac Island's place in railroad tourist literature. The book had sixty-one well illustrated pages, of which approximately five pages were devoted to Mackinac Island.[25] What was printed about the island in this, and other railroad publications, was usually a compilation of reprints. Mackinac Island's description included words from the "well-known author" Mr. Hallock regarding the island's scenery. It also included a ringing endorsement of the island's healthy climate from "the late Dr. Drake," whose long passage read in part, "The Island of Mackinac is the last and of the whole the most important summer resort to which we can direct the attention of the infirm and the fashionable."[26] The GR&I was the first, but certainly not the only, Michigan railroad to widely publish literature promoting tourism.[27]

Passenger ship companies had carried tourists to northern Michigan destinations long before railroads reached the same locations. In 1880,

Great Lakes passenger ships carried approximately 1.35 million passengers. By 1889, that number had increased to 2.23 million. By 1906, slightly more than 14 million passengers moved across the lakes' waters. Although the steamship business was growing rapidly, prudence suggested the companies that owned the ships begin publishing tourist literature to limit the possible impact of the railroads on the shipping lines' profits.[28] The first steamship advertisement aimed at tourists, a small four-page flyer, was printed in 1855, but for many years afterward the steamship companies largely ignored tourism.[29] Among the first steamship lines that printed tourist literature similar to that issued by the railroads was the Lake Superior Transit Company of Buffalo, New York, which in 1878 published *Attractions of an Excursion Up the Great Lakes*.[30] A few years later the Detroit & Cleveland Steam Navigation Company (D&C), which first scheduled a regular stop at Mackinac island in 1882,[31] began publishing tourist literature. The D&C's first clearly oriented tourist publication, *A Lake Tour to Picturesque Mackinac, Historical and Descriptive*, was published in 1884. Usually similar to the railroad literature in content, the D&C printed a publication, more or less annually, throughout the remaining years of the nineteenth century.[32]

The problem with the transportation company literature, from the island's viewpoint, was that while Mackinac Island received fulsome praise, the books also lavishly praised all the other places where the boats called and all the other places where the train stopped. The transportation companies' literature listed Mackinac Island as one of many desirable vacation spots. They did not distinguish Mackinac Island as a premier destination. A second set of travel books tried to address this concern. These books described one specific place in detail or were sold as guidebooks to those already there to help ensure the reader that they saw everything important.

The first such publication written about Mackinac Island was printed by John Disturnell. For much of the nineteenth century, Disturnell was a leading publisher of maps, travel guides, and directories describing North America, South America, and Europe. In 1875, he published *Island of Mackinac, Giving a Description of all the Objects of Interest and Places of Resort*. The book was so fulsome in its praise of Mackinac, one modern writer called it "almost a tribute" to the island.[33] Disturnell's unbounded praise for Mackinac continued the tradition of the pre–Civil War authors and was most often followed by subsequent travel guides. Disturnell, however, changed the content of books about the island. Like its pre–Civil War literary predecessors, Disturnell explained to people why going to Mackinac Island was worth the trip. Unlike the pre–Civil War books, Disturnell also told the potential traveler details about how to get to Mackinac Island and offered similarly detailed advice about where to stay and what to see.

Dwight Kelton's *Annals of Fort Mackinac*, first published in 1882, was also an important model for the subsequent island tourist literature.[34] Kelton had none of the publishing experience of Disturnell. He was an army lieutenant assigned at Fort Mackinac who took up writing "to help pass the long, dreary evenings of a Mackinac winter."[35] His inexperience made his work less polished than Disturnell's publication, but the book nevertheless distinguished itself in two ways. It was the first guide to the island written by someone who had lived there and thus could claim to offer the "real story." The second distinguishing characteristic of Kelton's work was a much greater emphasis on the island history than Disturnell or anyone else who previously published about the island had provided. Kelton included Native legends and many lengthy historic tables and lists that he personally compiled, such as historical chronologies, lists of past government officials connected to the island, and lists of the individuals married on the island.[36] Kelton's book was republished at least ten times before 1896, most often as a new edition appearing in the spring.[37]

Books like Kelton's and Disturnell's were helpful in singling out Mackinac Island before the public, but every tourist destination with aspirations of greatness was generating similar literature. As with the transportation company tourist literature, these books were not decisive in making Mackinac stand above other vacation destinations. With many places to choose from and extensive tourist literature that often made very similar claims for many locations, the public needed some additional way to decide where to take their vacation. For many, what helped make the decision was where public parks were located.[38] A park signified that the government recognized a site as being particularly noteworthy. Mackinac Island, in 1875, became home to the nation's second national park and the first national park east of the Mississippi River. If there had been a lottery to decide which resort would be crowned most scenic, east of the Mississippi, Mackinac Island would have been declared its winner. Being named a national park placed the island in its own category as a vacation destination.

How Mackinac Island was named a national park was a combination of beauty and good fortune. In 1872, a large area near Yellowstone Montana was designated as America's first national park.[39] On March 11, 1873, Michigan's US Senator Thomas Ferry, who through a stroke of good fortune for the island's future had been born on Mackinac Island,[40] proposed legislation directing the secretary of war to investigate creating a national park on Mackinac Island. The War Department was involved both because of the fort on the island, and because much of the island's land outside of the fort was a federal military reserve. Ferry wanted most of the military reserve to become a national park, excluding only Fort Mackinac itself and around one hundred acres of land surrounding the fort. Ferry's stated reason for his request was

> to hold intact, as far as practicable, this Island, fast becoming the favorite of the white man as well as the Indian, that I seek by the measure proposed to guard against its natural curiosities and beauty being lessened or destroyed by the hands of wanton despoilers.[41]

Ferry's concern about despoilers of his birthplace was not new. Worry that tourism would ruin the natural beauty of Mackinac Island was almost as old as the travel literature describing the island. With more tourists arriving each year, concern about their impact on the island's natural beauty had grown. What was new was that the concern over tourism despoiling places of great beauty had become a national issue. The example often cited for what could go wrong was Niagara Falls. Many were complaining that the Falls beauty was being ruined by the number of shops and the aggressiveness of guides and souvenir sellers. For example, Henry James wrote in 1871 that the Falls were:

> choked in the horribly vulgar shops and booths and catchpenny artifices which have pushed and elbowed to within the very spray of the falls. . . . The importunities one suffers here . . . from hackmen and photographers and vendors of gim-cracks, are simply hideous and infamous.[42]

Picturesque America, published in 1872, described Niagara Falls this way: "A superb diamond set in lead. The stone is perfect but the setting lamentably vile and destitute of beauty."[43] Public sentiment to preserve places of great natural beauty was growing. By the 1870s, some on the island were also concerned that commercialization was damaging the island's natural beauty.[44]

Ferry may also have been concerned that Congress itself might ruin the island. In 1871, Congress authorized the sale of "16 acres more or less" to Edward Franks, owner of the Mission House Hotel. The legal status of the land under and around Mission House had been unclear for decades. In 1825, the government had given permission to the American Board for Foreign Missions to erect a building on federal land for the purpose of educating Native American children. The Board had abandoned this project in 1837 and eventually disposed of the building as if it were private property. In 1845, Franks opened a hotel in the abandoned school.[45] But who owned the land beneath the hotel, and how much land that might be, was unclear. At least eight other island residents were in a similar situation, in some way encroaching on land owned by the federal government.[46] Ferry may have worried about what the consequences would be if these claimants, and who knows what other gim-crack vendors, persuaded Congress to give them a piece of the island. Achieving Ferry's goal of creating a national park on Mackinac Island was not easy, but President Ulysses S. Grant signed the legislation into law March 3, 1875.[47] Mackinac Island's claim to being a uniquely beautiful location was given a seal of approval by the federal government.

Establishing the park, however, proved easier than running the park. The federal government offered no money to operate it. This was not a surprise. Congress had also refused to pass an annual appropriation to fund operations at Yellowstone National Park.[48] Senator Ferry had likely anticipated this and likely assumed soldiers already stationed at Fort Mackinac could do most of the needed work, with the fort commander also serving as park superintendent.[49] That this plan ignored existing laws which made it illegal to spend military funds for park operations was both true and irrelevant. As Fery undoubtedly anticipated, the fort's commanders eventually found work-arounds to the law and did what was needed in the park.[50]

If the national park was the federal government's first gift to the island's tourist business, the authorization given to grant leases for cottages proved a second windfall for the island's image as a tourist attraction.[51] In 1885, the first three cottages were built on the island's east bluff. In 1886, the first cottage was constructed on the west bluff. By 1891, twenty-five cottages had been erected. Although called "cottages," these were not humble structures. Their owners were people of enormous means, who often made a public display of their wealth, and thus the wealth that had come to the island.[52]

Before cottages were constructed on federal land, Gurdon Hubbard had begun a similar project on island land he owned. In 1870, Hubbard had built one of the first summer cottages on the island. After losing most of his wealth in Chicago's Great Fire of 1871, Hubbard invested what money he had left into his property on the island, called Hubbard's Annex. In the 1870s, Hubbard began selling the lots to individuals who would build large "cottages" on each lot, while he promised to construct a similarly big hotel. By 1883, eight substantial cottages had been built by well-to-do Chicago residents, but no progress had been made on a hotel.[53]

Hubbard had made a beginning, but the cottages in the park set a new standard of both size and elegance. Mackinac Island "cottages" became ever grander mansions, with almost a competition among owners over whose cottage was the grandest. The motivation for building and building again was to keep up with the Joneses, or in the case of Mackinac Island, the Cudahys.

The Irish born Cudahy Brothers were business partners with Philip Armour in his meat packing business. They prospered first in the partnership with Armour and later as the owners of an independent company. By the 1920s, the Cudahys employed more than thirteen thousand workers nationwide.[54] The wealth generated by such a massive corporation gave the Cudahys the means to build on a grand scale, and soon a building race led owners of newly constructed cottages to enlarge their structure, or in a few cases, completely tear down their virtually new summer home for something even grander. All this building did not go unnoticed. In 1893, the *Petoskey Daily Resorter* wrote, "Handsome and elegant cottages are the only kind on Mackinac Island. They are the best in architecture and the richest in furnishings that wealth can command."[55] Without any planning or promotion, the island's cottages added to the sense a visitor

Mackinac Island Cottage of Michael Cudahy. Courtesy Clarke Historical Library, Central Michigan University.

had that Mackinac was the fashionable resort of many of the Midwest's most wealthy citizens. If the village remained rather unexceptional, the cottages made up for it.

What was needed to complement the extravagant cottages being built on the island's bluffs was a similarly extravagant hotel which could accommodate guests who expected to live as if they were the visiting the Cudahys.' But in the 1880s, none of the island's hotels and boarding houses could meet that standard. Three transportation companies built the sumptuous resort needed as the capstone of the aspiration to serve society's elite. In the process they not only gave the island the luxury hotel expected at a fashionable resort, but also gave themselves a reason to promote the island over and above the normal praise bestowed on all the sites a tourist might visit while traveling along the company's routes. The Detroit & Cleveland Navigation Company, the Grand Rapids & Indiana Railroad, and the Michigan Central Railroad built the Grand Hotel.

Within Michigan and across the nation, transportation company support for real estate speculation was not new. The Grand Rapids &

Indiana Railroad, for example, in 1876, had given approximately three hundred acres just north of Petoskey to the Michigan State Campground Association of the Methodist Episcopal Church, which founded the Bay View community.[56] The Grand Rapids & Indiana Railroad later created a Hiawatha pageant, a dramatization of Longfellow's epic poem, *The Song of Hiawatha.* To create a home for the pageant, the railroad built a new park about three miles from Petoskey. To take tourists to the park, a new train station and additional track was built. This new train station served no purpose other than to bring tourists to see the play.[57]

Knowing that transportation companies often engaged in real estate speculation, in 1886, John Oliver Plank, an experienced hotel owner and manager, pitched the idea of a new resort hotel on the island to officials from the three companies. He dismissed the then existing accommodations on the island as ones that could "hardly be dignified with the word 'hotel.'" Plank wanted nothing less than the best, even though as he recalled, "people laughed at him for proposing such a large building in such a remote place."[58]

What Plank had in mind likely was modeled on beach hotels then existing all over the United States,[59] all of which could trace their design to luxury hotels in the Catskill Mountains, particularly the Catskill Mountain House which had opened in 1823. That hotel was built at the edge of the Catskill Mountains, with a magnificent view. To better enjoy the view, the hotel featured a long, colonnaded porch. The Mountain House was also luxurious. Even European visitors, who seemed to take perverse satisfaction in complaining about the accommodations they found in America, agreed the Mountain House offered sumptuous living and dining equal to any fine European hotel.[60]

How Plank persuaded the representatives of the three companies to not laugh at his vision for the new hotel similar to Mountain House was never truly explained. *Hotel World* described what happened this way: "The Grand was conceived in the brain of Plank, who with the meekness and simplicity of a shepherd leading his flock, led three great transportation companies to invest over $250,000 in a large wooden structure."[61] The transportation companies became equal partners in the Mackinac Island Hotel Company. Plank, in addition to obtaining shares in the company, leased and managed the facility, which, to the surprise

Grand Hotel, ca. 1900. Detroit Publishing Co. Courtesy the Library of Congress.

of the three transportation companies, was named not the Grand Hotel but "Plank's Grand Hotel."[62]

Plank's Grand Hotel opened on July 10, 1887.[63] Inside, the hotel was uninspired. The interior was uninspired. What made the hotel stand out was its porch. A later manager of the Grand, not happy with all of hotel's construction details, admitted that all the errors were compensated for by that "magnificent porch." The porch had an unparalleled view of the water. It was also an ideal place to promenade, a pastime Victorians took great delight in doing. After lunch guests would dress in fine clothes and slowly make their way up and down the porch, stopping frequently to greet similarly attired friends.[64] On some evenings, after a fine dinner, a formal dance was held. Even more than during the afternoon promenade, the evening dance was the place to see and be seen. Men wore formal attire and women appeared in their finest gowns. To take in the spectacle, the

Grand's dance floor was surrounded by decorative railings and box seats. There was also a balcony from which to view the dancers.[65]

Plank had built what the island needed to claim to be Michigan's resort for society's elite. A Cheboygan newspaper underscored this when it remarked that the project would bring a "better class of people than have ever before visited the Island."[66]

With outstanding natural beauty, a reputation for promoting good health, the first national park east of the Mississippi River, grand "cottages," a downtown that was beginning to revitalize,[67] and a destination hotel that claimed to be where the elite from Chicago and the Midwest came, Mackinac Island had become "among the largest and most fashionable midwestern resorts."[68] The future for tourism on the island appeared bright, with the island emerging as a premier Great Lakes destination. But that brightness soon was cast into shadow. On October 9, 1894, the soldiers marched from the fort to the island dock for the final time. Congress had demanded new economies from the military and the army had closed Fort Mackinac.[69] Without the soldiers to maintain it, island residents faced the grim prospect of what would happen to the national park and the tourism it generated.

CHAPTER 3

Making the People's Resort, 1895–1930

The closing of Fort Mackinac created a crisis. Without soldiers the national park lacked both management and workers. For all practical purposes, the park was closed. Without the park the island lost its claim as a uniquely beautiful locale. One of the principal underpinnings of the local tourist industry was threatened. To stop this threat, the island government sought the help of James McMillan, one of the state's two US senators.[1] A resolution was also introduced into the Michigan Senate directing Michigan's congressional delegation to work to ensure that the federal government continued operating the park.[2]

The efforts by both local government and the state legislature had no impact. Senator McMillan knew there was no chance the army would reverse its decision. Thus, in February 1895, McMillan introduced legislation that required the federal government to transfer to Michigan's government all federal assets on the island, if the state asked for them. Michigan could affirm the uniqueness of the island's natural beauty by making it a state park. There was precedent for this. In 1864, the federal government had given to the state of California much of the Yosemite Valley for use as a park.[3] McMillan's bill became law.[4]

The problem was that unlike California in 1864, where the legislature was eager to create a state park from the land given them by the federal government,[5] Michigan's elected officials were not interested in establishing what would be Michigan's first state park. Their lack of interest most likely came from an unwillingness to spend taxpayer dollars on something they saw as frivolous. McMillan undoubtedly knew this when he sent Michigan Governor John Rich a pointed telegram that both addressed the urgency of the matter and suggested a way to pay for the park without a state appropriation. Regarding finances, McMillan wrote, "State can . . . maintain park as New York maintains Niagara Park. Expense can be supported by receipts from leases." Lest the governor believe he could take the time to think about the matter, McMillan warned him in the same telegram, "Senate appropriations committee is considering seriously the sale of Mackinac Island Park to private parties."[6] McMillan's warning about private parties was most likely intended to invoke the unhappy history of Niagara Falls, where much of the site's natural beauty had been defaced by overly aggressive commercial development.[7] Faced with the possibility that the fort and park would be privately purchased and perhaps poorly used, and given a theoretical way to pay for the park's operation with something other than an annual appropriation, the government in Lansing accepted the federal government's land on Mackinac Island. All the necessary legal steps were completed before the summer ended.[8] Michigan had created its first state park. The question was what to do with it.

On May 31, 1895, Michigan created the Mackinac Island State Park Commission to operate the park.[9] The commission was composed of five commissioners, each of whom was appointed for a fixed ten-year term in office. As originally introduced in the state senate, the bill creating the commission had language within it calling for regular appropriations to fund the park. When it considered the bill, the Michigan House of Representatives deleted all references to money. When the revised bill was returned to the Senate, the House changes were unanimously accepted.[10] The five commissioners were told to run the park with no state money. This obviously was a problem, but, perhaps even worse, the law establishing the commission was silent regarding the body's objectives. The legislature left it to the commission to decide what purpose the park would serve.

In 1895, the park's purpose took a back seat to the newly appointed commissioners immediate need for money to run the park. At its first meeting, held July 11, 1895, the Mackinac Island State Park Commission tried to raise some of the money it needed by doubling the rent of cottage owners.[11] Collecting the increased rent proved challenging,[12] but even if it had all been quickly paid, the money raised fell well short of what was needed. Knowing this, the commission implemented other fundraising schemes. The commission granted new commercial concessions in the park, for example establishing one at Arch Rock for a two-hundred-dollar annual fee.[13] They tried growing and selling vegetables in the old garden where soldiers had grown vegetables. In 1898, the commission allowed cattle to graze on park land all season, for a five-dollar-per-cow fee.[14]

Income was so slim the first year that there was only enough money to pay one employee, a watchman. George T. Arnold, the only commissioner resident on the island, was pressed into service as park superintendent without pay. The next year, in 1896, the commission scraped together funding to pay a park superintendent and three watchmen. But in September 1897, because of ongoing funding problems, the paid park superintendent was laid off. Arnold found himself again serving as superintendent, again without pay.[15]

All this was a far cry from the ninety to one hundred soldiers that had previously cared for the park.[16] The commission's failure to adequately replace the army led to growing dissatisfaction over the park's condition. In February 1897, the Common Council of Mackinac Island expressed the collective unhappiness of area residents by sending to the legislature, with the council's approval, a petition signed by 109 citizens of Mackinac County asking that the state give the park back to the federal government. The Michigan legislature, which had not really wanted the park in the first place, unanimously passed a bill requesting that the federal government take back ownership and operation of the park and fort.[17] In the nation's capital, the legislature's resolution was received and filed without any action.

Likely knowing this would happen, the same session of the state legislature that asked the federal government to take back the land also considered a bill to pay for the construction of a beach front road to circle the island. The road would help the commission raise additional income, by having available more waterfront lots to lease for cottages. The road bill had considerable support in the state senate, where it passed easily. In the state house, however, it generated, in the words of the official record, "considerable discussion." It eventually passed the House 61 to 26.[18]

Governor Hazen Pingree vetoed the bill. In his veto message, Pingree did not discuss if the proposed road was a good idea. Rather, he wrote, "I understand that there is sufficient revenue from the rental of lots to take care of the public property now upon Mackinaw [*sic*] Island." The commission might have accepted Pingree's veto with little disagreement if he had simply said that in hard times there was no money available for the road, something he also discussed in his veto message.[19] What stood out, however, was Pingree's assertion that the park commission already had enough money, something that the commissioners knew was not true.

The bill's fate set the pattern for how Lansing would treat its island park—as an unwanted millstone. Eventually, beginning in 1904, financial reality led to the passage, in most years, of appropriations for the park commission.[20] But it was usually a small sum, short of what the park commission wanted, and most often needed.

If the state was negligent in supporting its island park, Lansing completely failed to direct the commission regarding the mission and purpose of the park. With no legislatively established goal, the commission, in 1900, unilaterally defined why the park existed and who it would serve in their mission statement:

> Mackinac Island is the most beautiful and healthful resort in the Northwest. It is becoming more the place where the tired teacher, clerk, housekeeper, businessman and farmer go to spend their short vacation of a few days.... While the rich can build their palatial cottages on the Island, the people of moderate means are making it more and more their playground and resting spot. The commission desires to make this park and Island more emphatically "the people's resort."[21]

The embrace of the island's natural beauty and the declaration that that beauty should serve as a backdrop for "the people's resort" was rooted in trends sweeping the nation at the time.

After the Civil War, a deep divide developed over what should be the purpose of public parks. As the middle class and, increasingly, less affluent day-trippers, began to join the rich in taking vacations, it became apparent that these groups had very different expectations for their trip. The rich, who had dominated travel prior to the Civil War, tended to see a vacation in terms of idyllic strolls along picturesque landscapes. The day-trippers in particular might take in the scenery but also were quick to visit the diversions that often were associated with major tourist destinations and often offended the sensibilities of the well-to-do. Niagara Falls, because of its prominence, became one of the first battlegrounds between these competing views of vacationing.

In 1883, society's elite persuaded the New York legislature to create a park to push back the carnival-like attractions that had crept up to the Niagara River's edge. Frederick Olmstead, who designed the new park, consciously sought to restore the original picturesque beauty of the Falls landscape. When Olmstead was reminded of the less affluent's enjoyment of the carnival attractions, he dismissed them. The interest in such tawdry things was the product of poor taste combined with being led astray by commercial vendors. Olmstead added that if the park were built as he designed it, Niagara Falls would not only be restored to its original beauty, that restored beauty would help civilize the day visitors. New York built Olmstead's picturesque park, but it did not civilize the masses. The carnival-like vendors continued to be well patronized after they set up new shops just outside the new park's boundaries.[22]

On Mackinac Island, there was both a picturesque landscape and a town that thrived through tourist attractions. Unlike Frederick Olmstead, however, the Mackinac Island State Park Commission was conscious of and interested in the opinions of the less affluent. The commission accepted the existence of vendors within some parts of the park and within the island's village, on land not under the commission's jurisdiction. The arrangement was not ideal, since it created a significant tension between the interests of a picturesque park and the sometimes carnival-like atmosphere created by privately operated tourist services and village merchants, but it realistically reflected the multifaceted character of the park's various users.

That the first individuals appointed to the Mackinac Island State Park Commission embraced such a broad view of the park's audience is something of a surprise. The men the governor picked to serve as the first park commissioners were people of significant wealth. Given their background, it would have been likely that they would reserve and protect Mackinac's picturesque character, and by extension its role as an enclave for the wealthy, over the middle- and working-class visitor.

Of the four original members of the commission for whom biographical data can be found, only George Arnold was not a man of extraordinary wealth. He was, however, a very successful businessman who had moved to Mackinac Island from southwestern Michigan in 1881. Arnold invested widely in the region and was called the "chief entrepreneur" of the Straits of Mackinac, owning among other things two island hotels and a straits-based ferry company.[23] He certainly had the most to gain by promoting commercialization, but he was the lone voice on the commission with an economic interest in the island's tourist trade.

Thomas Ferry was also appointed to the commission. He had inherited a fortune made in the lumber business, but when he left the US Senate in 1883, much of his money was gone. Nevertheless, he was still very comfortable.[24] Commissioner Albert Stephens was considered the state's most successful lumber baron. When he visited Mackinac, he likely arrived on his 152-foot yacht, *Zara*.[25] Peter White was also very rich, making his fortune in iron and then investing what he had made with great success. Among the original commissioners he was the only person with a reputation as a philanthropist, interested in the lives of people from the middle class and working class.[26] The historical record is silent regarding who drafted the commission's mission statement and what discussion preceded its passage, but it seems probable that Arnold and White joined forces to persuade their fellow commissioners to adopt a broader view of who the park would serve.

How the island's business community reacted to the commission's mission statement is also not well documented. On the surface, some of the island's business community could be assumed to want to have the park that served as a retreat for the wealthy. The Grand Hotel, in particular, may have wished this since it was catering to, in the words of a Cheboygan newspaper, a "better class of people than have ever before visited the Island."[27] However there is reason to suspect that even the owners and managers of the Grand Hotel were quietly pleased by the commission's mission statement.

The owners of the Grand Hotel were three transportation companies which made money selling passenger tickets to anyone with the means to buy them. Yet the hotel was imagined by John Oliver Plank not as a destination for the railroad and boat line passengers, but rather as Michigan's premier resort for society's elite.[28] It was a business model that proved a financial failure. Three years after it opened, the Grand Hotel was in financial trouble. It was not attracting a sufficient number of guests willing to pay a premium to stay there. In the fall of 1889 hotel publications began to hint at or just say that Plank was overcharging his guests and mismanaging the operation. Plank blamed the owners, who he claimed had not invested enough money in the hotel. Whatever was the truth, by the fall of 1889, both Plank and the hotel's owners were tired of losing money and were ready to part company.[29]

That changes in the Grand Hotel's management were coming was public knowledge. The *Petoskey Independent Democrat* wrote on April 27, 1890, "The owners of the Grand Hotel are trying to get [James R.] Hayes to take charge of their big white elephant."[30] On May 20, 1890, Hayes became manager of the big white elephant on Mackinac Island while also managing two other hotels, the Arlington in Petoskey and the Wayne Hotel in Detroit.[31]

Hayes was a personable man who charmed his guests. To his personality he added a shrewd business sense and even shrewder marketing skills. Admitting by his actions if not his words that Plank's hope for guests who would pay a premium price for rooms had failed, Hayes quickly, if quietly, lowered the daily room rate from four dollars to three dollars. He likely had to because the Grand Hotel's competitors, such as Astor House, were only charging ten to twelve dollars a week for a room. To better entertain guests, Hayes convinced the hotel's owners to construct a new "activity casino" or perhaps, since it was something Plank had wanted but the owners refused to spend money on, made its construction a condition of his accepting the job. Whatever the case, the casino was built offering new recreational possibilities.[32]

Plank's Grand Hotel had a proper Victorian feel to it, as one would expect of a resort designed for society's elite. However, the days of the upper-class summer hotel were waning. It was held by many to be a world of rich women who cultivated the arts of fancy dress and leisurely conversation. Men often found it boring, but so too did a growing number of women.[33] Hayes realized this and devised a program that was aimed at the middle class, by making sure people at the hotel always had something to do. One thing to do was dancing, for which Mackinac Island already had quite a reputation. As Angeline Teal wrote in 1883,

> Mackinac is well supplied with spacious and elegant hotels, filled during the season with summer tourists. They dance a great deal there. At Petoskey it is different. The near proximity of several resorts which are under the influence of religious associations, and the character of the multitude of those who flock to these resorts during the camp-meeting season, has left its imprint on the entire social life of the place. He who, in the language of a gay wordling [*sic*] of my acquaintance, likes "a very little Sunday-school and a good deal of hop" in his summer recreations, will naturally prefer Mackinac.[34]

On Saturday night Hayes continued to host a formal dance like that Plank had sponsored. But Hayes added two additional dances during the week that were "informal and the boys [could] skip around in their cowhides if they want." To make "a good deal of hop" even more popular, Hayes hired a dancing instructor.[35]

Hayes further entertained his guests with all kinds of sporting activity. Tennis, bowling, and boxing were all to be found on the grounds. He arranged boat races that ended at the hotel. Thirty horses and an equestrian instructor were available. Hayes adored bicycles and made sure guests could use one if they wished. For those who preferred less strenuous sports, he invented nonsensical competitions designed as pleasant amusement. One was a foot race involving one match, one cigar, and one umbrella, none of which could be placed on the ground. The winner was the first man across the finish line with a lit cigar, and still holding the umbrella and the match.

Hayes's greatest invention was the annual Field Day. Held in August, it featured everything from real sports to novelties. A favorite novelty was the Grand's version of a greased pole contest. In the hotel's rendition, a ten-dollar bill was placed at the end of a twenty-five-foot pole that was extended horizontally out over the lake. The money was anyone's for the taking, if the guest could get to it. A few succeeded. Most who took the challenge got dunked in the cold lake water. Watching young men struggle toward the ten-dollar bill was easily the day's most popular entertainment. Field Day was for years the Grand's busiest weekend and most important social event.[36] Field Day was also a very early example of what would in the future become a standard tool to attract tourists, the multiday festival.

When the music stopped and the games ended, Hayes offered other things to interest his guests. In 1895, Mark Twain spoke at the Grand.[37] Hayes purchased one of Thomas Edison's phonographs as soon as he could get one and demonstrated the new invention on the hotel's porch to his amazed guests.[38] Another of Hayes's innovations was a chalkboard. Beginning in 1894 each day Hayes had posted on the chalkboard the expected high temperature on Mackinac Island, as well as the expected high temperatures in major midwestern cities. Guests, comparing the pleasant temperature forecast for the island compared to the much hotter forecasts for the various cities, would invariably announce, "Aren't we lucky to be where we are!"[39]

Hayes's greatest concession to the middle class was to seek out the convention trade. There was no publicity about this part of the business, but in 1898, the Detroit papers were talking of the Grand as having "quite a reputation as a convention place."[40] Bringing conventions of businessmen and fraternal organizations, along with his other innovations, filled rooms with a middle-class clientele. It was a very good way to make money, but Hayes never talked about it, at least not anywhere where a newspaper reporter could quote him on the subject.

To the press he consistently presented his hotel as the place where the "elite of Society played."[41] To accomplish this, he made sure the Chicago papers were regularly informed of prominent guests from the southern shore of Lake Michigan. The *Chicago Sunday Herald* regularly printed the names of the important Chicagoans actually, or supposedly, visiting the Grand. The ambiguity was caused by Hayes's practice of including on his list people who he said had made plans to come, rather than only mentioning the names of actual guests.[42] Hayes also supported fledging publications like the *Chicago 400*, avidly read by social climbers in the city, as well as emerging tourism magazines, advertising in them all with the implicit and perhaps explicit expectation that they would return his financial support with kind printed words about Mackinac Island generally and the Grand Hotel in particular. Although he targeted Chicago, Hays made sure the Detroit papers carried similar stories about the important members of that city's society visiting, or planning to visit, the Grand.[43]

Hayes presented an image of the Grand Hotel as a place where society's elite stayed while he made his money by selling rooms to the middle class. As a result, he was likely not truly disturbed about the commission's decision to declare the new state park one for the people since the people the park commission had in mind were the same ones filling his rooms.

Others who made their money openly from middle-class travelers pushed the commission for greater commercialization of park land. One such group were the investors who in 1898 asked the commission to lease the land immediately below the fort to build a new hotel. This was prime land facing the water, and the lease would have generated considerable income for the cash-strapped commission. However, some people, most prominently an organization named the Marquette Monument Society, wanted the property reserved for a landscaped park which would feature a statue dedicated to Father Jacques Marquette.[44] The commissioners voted in favor of using the land for a park. To make their decision more emphatic they held the commission's first roll call vote, unanimously setting aside the land to be "a park for monumental and ornamental purposes."[45] It was a critical precedent that meant island commerce would have to be measured against public good. In 1905, private funding made it possible to begin the park's landscaping and a privately funded statue of Father Marquette was unveiled in 1909.[46]

The commission took very seriously its self-created mission to establish and maintain a park for the benefit of the public. It also took very seriously a second part of the 1900 mission statement. When it included the phrase "this park and Island" in its mission statement the commission subtly but with apparent intentionality included within its purview the island's private property. The commissioners assumed a say inside the island's village. This self-made grant of authority would lead to endless, and sometimes bitter, arguments. The arguments began quickly, if in a small way. Island residents wanted modern conveniences like electricity and telephone service. The commission often found the means employed to accomplish modernization aesthetically undesirable. Telephone poles

Marquette Park, ca. 1906 to 1908. Detroit Publishing Co. COURTESY THE LIBRARY OF CONGRESS.

were an early point of conflict. The Mackinac Island Electric Light and Telephone Company repeatedly placed poles in locations the commission disliked. The commission repeatedly asked, usually successfully, for poles to be relocated.[47]

The locations of telephone poles were, in truth, a minor matter. Eventually however, the commission sought to regulate private business. Within reason, this was necessary. Island businesses had a reputation for excess. The problem was neatly summed up by a phrase used to describe some local merchants: "human sharks."[48] A 1919 newspaper article expanded on that moniker with the following unflattering description:

> Extortionate prices have made it impossible for persons of moderate means to spend even a few weeks there [Mackinac Island]. The Island has been

> turned into a hack drivers' heaven. The cabmen are practically the only permanent residents and elect themselves to positions in local government then vote themselves good salaries.
>
> The steamships [ferries] running to the island pay no attention to agreements carrying passengers and mail, or to the convenience of the public. If the number of passengers on hand does not assure a profit for any given trip, the owners are likely to cancel the sailing. They take the boats off in the fall before the time agreed on.
>
> Private owners of the beautiful forest lands on the Island are timbering them and leaving unsightly stretches of waste along the main drive. Land speculators are now trying to close one of the most picturesque roads.[49]

"Timbering off" island land was a practice of cutting down all of the trees within a given area. As for the "hack drivers," the island's carriage operators would become the island business group most often in conflict with the commission. Commissioners complained regularly about the island's carriage operators and passed numerous rules and regulations to control them, using access to the park as a club. None of the regulations were welcomed by the carriage operators and none worked in a way satisfactory to the commission. The two groups saw the world very differently, with each side convinced about the virtue of their point of view.[50]

The carriage operators had a long history of rowdy behavior when competing for customers. In the 1920s, one person recalled that when a passenger ship arrived "it was quite a show to see the carriage operators swarming up the gang plank to latch onto unprepared tourists and promise them the best views of the island from their carriage. It became such a problem that ships' officers finally forbade any carriage owner from coming aboard."[51]

In addition to rowdiness, some tour operators began to charge fees that the commission considered exorbitant. The operators added insult to injury by not only charging more but shortening the length of their tours. In the 1920s, the commission wanted both fees and tours standardized. Carriage tour operators believed their fees and tour routes were none of the commission's business. Tradition was on the tour operators' side. Carriage tours of the island's sites had been offered to visitors since the mid-nineteenth century. The liverymen had never had a fixed price for tours but rather always charged what the market would bear. And while technically the carriage operator hired to give an island tour would take their passengers wherever the guest wanted to go, most passengers left the route to the liveryman, who could make the ride as long, or as short, as he wanted.[52]

The commission was, however, uninterested in the history of the carriage trade. Unless things changed, the commission threatened to enact rules regulating carriage tours within the park that most carriage operators greatly disliked. To avoid formal regulation, some operators proposed a self-regulating carriage association that would create uniform fees and tour practices acceptable to the commission. Feelings about the proposed association ran so high among the carriage operators that fistfights between drivers broke out on the streets.[53] Despite the brawls, the drivers finally organized themselves into single association, which demanded of the commission more than it had originally offered. In return for their "voluntary" self-regulation, the commission granted the association members the exclusive rights to run tours in the park and make use of park roads for taxi services.[54]

The carriage operators' loose practices that were so concerning to the commission were nothing compared to the loose practices at the Grand Hotel during the 1920s. The island's leading business was being run as a gambling casino and a speakeasy. Illegal activity, particularly gambling, was not new to the Grand Hotel, but what was new was both its scale and public acceptance.

The heart of the problem was the Grand Hotel, which despite all of Hayes's efforts and those of the managers who followed him, had never earned a reliable profit. Eventually the hotel's original owners sold their interest to Henry Weaver, who had taken over operation of the hotel in 1899.[55] At the end of the 1910 season, Weaver said he was considering closing the Grand for good, recovering what money he could by selling the building for scrap lumber.[56] To avoid this, three island residents, Charles Holden, who in the 1910 federal census was listed as owning an island saloon, George Arnold, owner of Arnold Ferry Line and a park commissioner from 1895 to 1903, and Patrick Doud, a local contractor, became stockholders in the hotel. Charles Ballard of Indiana may also have been involved in the matter.[57] While this kept the hotel open, it did nothing to solve the hotel's chronic financial problems. Whether or not the Ballard family was involved in earlier financial transactions, when the 1918 season began, James Logan Ballard, Charles's brother, was listed as the hotel's owner.[58]

The Ballard family knew how to turn a profit, although not a legal one. In 1894, Charles Ballard had taken over a saloon in French Lick, Indiana, a popular tourist destination. Charles improved the bar's profit substantially by adding illegal gambling. Charles Ballard soon was organizing illegal gambling for several French Lick establishments. By 1915, Ballard was operating "one of the most famous gambling establishments in the Mid-West." He would eventually become owner of West Baden Springs Hotel, a palatial establishment of a size and character much like the Mackinac Island's Grand Hotel.[59]

Reports from as early as 1911 suggested the Ballard family were making money through gambling at the Grand Hotel. But the initial gambling operation was apparently small. When James Logan Ballard became the Grand Hotel's sole owner, that changed. The Ballards systematically applied what they had learned at French Lick to grow their Mackinac Island business.[60]

Alcohol was a second revenue stream for the Ballards. It was a lucrative money-making opportunity made possible by laws outlawing demon rum that were first adopted by Michigan and subsequently the federal government. In 1887, Michigan enacted a law allowing the state's counties to ban the sales of alcoholic beverages within their borders. Known as "local option," in 1907, only one county in the state was "dry," that is, had local option in force. However, public opinion in Michigan, and nationwide, regarding the sale of alcohol was rapidly changing. In 1910, voters in Mackinac County, which included Mackinac Island, made the county dry. In November 1916, Michigan voters passed a referendum making the entire state dry beginning May 1, 1918. On January 17, 1920, Michigan's dry laws were superseded by federal statute when Prohibition became the law throughout America. Like many others, Ballard made a great deal of money by ignoring the law and selling a drink to anyone who wanted one.[61]

During the 1920s, in the books the Ballard's presented publicly, the Grand Hotel was still losing money. Officially the hotel was a half-million dollars in debt.[62] But the public books did not, could not, include income from gambling and the sale of alcohol. The real balance sheet made clear that the Grand was turning a substantial profit.

The park commission, with its mission to create a wholesome island tourist experience for the tired teacher, clerk, housekeeper, businessman, and farmer likely found the Ballards nothing short of scandalous. A more tolerant commission would have realized that gambling had been an integral part of the success of most East Coast resort hotels, thus its appearance at the Grand was, if not welcomed, predictable.[63] But the commissioners were not tolerant of public vice, particularly when it was as public as the Ballard's operation in the Grand Hotel. What they found equally troubling was that the island community supported what the Ballards were doing.

In August 1919, the Grand Hotel was raided by the state police because of illegal gambling.[64] There is little doubt that the park commissioners thought it was high time to call the Ballards to a legal accounting. What likely surprised the commissioners, however, was the community reaction. The Mackinac County prosecutor, who had jurisdiction over the case, was unwilling to file criminal charges against the hotel's visitors or owners.[65] The island's mayor was furious about the "invasion" of state officials. The *Detroit Free Press* summed up the situation, reporting, "Inasmuch as practically every resident of Mackinac Island is favorable to the continuance of the gaming, it may be that no arrests will be made."[66] Gambling was just another part of the tourist business. Indeed, such was public opinion that Ballard filed court papers demanding the return of approximately $30,000 worth of seized gambling paraphernalia, claiming that since the crimes alleged were misdemeanors rather than felonies, the items could not be legally seized.[67]

Widespread gambling was again drawing tourists to the hotel in 1920, when another raid took place. This time, Ballard was arrested for both running an illegal gaming operation and selling liquor, although, as in 1919 no hotel guests were inconvenienced with an arrest.[68] State police again raided the hotel in 1921. Despite the raids, Ballard never seemed to have suffered any significant legal consequences nor lost any tourist business. Cynical residents noted that all of the raids had taken place late in the season, when many guests had already left.[69] The rumor was that the raids at the Grand were actually requested by the Ballards, who welcomed occasional newspaper stories regarding police raids as publicity, which slyly advertised activities available at the Grand, or as the locals put it, a little "spicey" publicity to help business.[70]

To address everything from brazen lawlessness to bad attitudes among Islanders, the commission adopted the radical solution that it should take control of the entire island by purchasing or otherwise seizing all private property. The commission had to address some property issues. Among those things gifted to the state by the federal government were a number of unresolved property disputes. In addition to these historical problems, some local businesses had played fast and loose with property lines. When the Chippewa Hotel was built in 1902, one corner of the building "strayed" onto park land. "Strayed" was a word allowing for the possibility an accident; the decision by the manager of the Grand Hotel to have an existing stone property marker moved so that a new fence would appear legal was a blatant act of trespassing. The manager laughed off accusations that he had impinged on state property and got away with it for many years, even though surveys in 1905 and 1913 confirmed the violation.[71]

The commission acted in the public interest when it opposed the illegal privatization of public land, but the commission began to act more questionably, based both on its mission statement and on changing thoughts about what a park like the one it administered should be. The National Park Service, created in 1915, had begun to argue that the true purpose of a national park was primarily to preserve important landscapes. To do this, the Park Service developed a three-part plan, allowing auto access to some parts of the nation's national parks, setting aside some part of parks for either hiking or horseback riding, and preserving some parts of parks by essentially making them inaccessible. Regarding the balance between preservation and tourist attractions, in 1922, the head of the newly created National Park Association, a private body that supported the Park Service and raised funds for park projects, said:

> They [national parks] embodied in actual reality and in splendor the American-born ideal of nature conservation, creating and protecting by law a

> mighty system of national museums of the primitive American wilderness was an inspiring offset to the sordid commercialism and much advertised political rottenness of the times.[72]

Whether their purpose was to rein in excessive commercialism or reflect the newly articulated view that parks such as the one on Mackinac Island be protected from any form of commercial activity, the park commission began to aggressively seek to obtain private land. One celebrated example was a fight the park commission picked with county commissioner and businessman James Doud. The disagreement was over a once privately owned forty-acre parcel of forested land. When the land became available for purchase, the park commission wanted to buy it, but, as was so often the case, it had no money to spend. In the end, the parcel was split. Doud bought twenty-three acres of the land while the rest was bought by a park commissioner who promised to keep it off the market until the commission could raise the money to purchase it from him, at the same price he had paid for it.[73]

In 1920, the park commission bought the land that was being held by its public-spirited member. Doud had annoyed the commission by cutting down trees on his part of the land, tree removal being something which the commission routinely opposed. Doud, who was planning to create a pasture and a fruit orchard on his land, cared not at all about what the commission thought of his activity since, he was removing, what was in his opinion, "the jungle that is now upon the land."[74]

To protect the park's land from Doud's axe, in 1922, Park Superintendent Frank Kenyon ordered the erection of a fence between the two parcels. The fence Kenyon planned generously defined the park's land holding and in all probability intentionally strayed into Doud's property. Kenyon's intent to use the fence to grab as much of the land as he could was no secret. When builders began constructing the fence, Doud ran them off the property one day, and when they returned the next day, used a loaded gun to persuade the work crew to go home.[75] Public sentiment was firmly on Doud's side, and, reflecting this, local law enforcement refused to take any action against Doud. The state attorney general eventually ordered Doud arrested and tried for felonious assault with a revolver. Doud was convicted, in a trial the *St. Ignace Enterprise* declared a "farce."[76] The trial did nothing to resolve the property dispute, which dragged on until 1930.[77]

The Doud dispute was a small matter compared to what the commission wished to achieve. After World War I, Michigan's governor was considering creating a state park system. As part of this idea, the Mackinac Island Park Commission suggested the state purchase all of the private land on the island and give it to the park commission. It was reported that the governor favored the idea.[78] Although that proposal failed, no one could miss the point when in 1923 the usually tight-fisted legislature gave the park commission $90,000 in one time money to purchase a six-hundred-acre island farm. With the purchase, the state took possession of about one-half of the island's remaining private land.[79]

Purchasing land, however, would likely never get the commission what it wanted. Some landowners might never sell. To remedy this problem, the commission asked the Courts to enforce controversial, and even if valid, long unenforced, land claims. The most dramatic battle took place over a large swath of village beach front. The commission alleged federal records gave it ownership of about one hundred acres of land, beginning at the village waterfront and moving inward through the town. The title to this land was unclear, but for more than fifty years, buildings had been constructed on it and the building owners paid property taxes annually to the city.[80] Among the structures on the land were the Chippewa and Iroquois hotels, approximately forty commercial storefronts and perhaps twenty-five cottages. This construction, the commission argued, was a mistake. The commission asked the court to give it the title to the land and shared its plan to rectify the error by removing the existing structures.[81]

The commission also used a legal claim to threaten the Grand Hotel. It began a legal dispute with the city of Mackinac Island and the Grand Hotel over the "Burough" lot. The city had been granted the property by the federal government and it had, in turn, given the Grand Hotel a ninety-nine year lease to it as an inducement to build the structure. However, within the agreement between the city and the federal government was a clause saying that the land could be used for "public purposes only." The commission argued that the city had violated this clause because the Grand Hotel was a private entity and thus its use of the land was not a "public purpose." The lease, the commission reasoned, had been violated and because of that control of the land should revert to the federal government's legal successor, the park commission.[82] One can only imagine the reaction of the Ballard family when they considered the possibility that some of the land under their hotel could become owned by the park commission.

The lengthy litigations over the Burough lot and over the beach front property made the commission many long-lived enemies within the community. In the end, the commission lost both the Burough lot case and the beachfront property case.[83] Even pistol waving James Doud had his original conviction overturned and was given a new trial.[84] The commission painfully learned it could not gain legal control over all the island's private property either through purchase or court action. But before the lesson was finally learned, many people realized that the bitter feuding between the business community and the commission was doing no one any good.

Tourism was the business of the island, and it was the mission of the state park commission. Open warfare regarding who controlled the land diverted energy from the development of tourism. Some kind of understanding was needed. The hope for an understanding was facilitated by the sudden death of James Logan Ballard. The immediate aftermath of Ballard's death only reminded everyone of the yawning gap between the commission and the community. The local press wrote, "Mr. Ballard was one of the best friends the Island had ever had spending money lavishly to further beautify and make popular the resort, and was always boosting and urging others." During the hour of his funeral, "Mackinac Island people showed their respect for him . . . by suspending all business, the tolling of church bells, and placing flags at half-mast."[85]

If the park commission likely did not lower its flags to half-mast in remembrance of a man it viewed as the island's leading lawbreaker, the commissioners could hope that change might be coming. The Ballard family initially discussed selling the Grand Hotel, and management temporarily fell into the hands of the hotel's president E. J. LaChance, a life-long resident, and W. Stewart Woodfill, the institution's young general manager. The possibility existed that the island's best known corporate lawbreaker might become, if not a model corporation, at least less flagrant in its lawbreaking.

As a start toward a more cooperative future, in August 1923, approximately three months after Ballard's death, the Mackinac Island Civic Association was created. This group brought together virtually everyone with a stake in tourism: the city council, businessmen, the cottagers' association, the park commission, villagers, the transportation companies, and the hotel and boarding house operators. Officially it was "a clearinghouse of ideas for the preservation and advancement of Mackinac Island." It was also a place where ideas could be discussed and resolved before becoming court cases. Or, as a newspaper story had it, "they can be dispensed with without hurting anybody's feelings."[86] Offering even more hope of some kind of reconciliation between groups, the association's first leader was E. J. LaChance, president of the

Mackinac Island Main Street, ca. 1915. Detroit Publishing Co.
COURTESY THE LIBRARY OF CONGRESS.

Grand Hotel. The Grand Hotel hosted a banquet celebrating the group's creation.[87] As the island's warring factions broke bread, the gathering signified the hope for a less combative future.

And there was agreement on many points. The old island claims of good health and beauty were things everyone agreed on and agreed to promote. The Mackinac Island Civic Association's 1926 tourist publication boasted that the island had "NO HAY FEVER, no malaria, no mosquitoes or other insect pests, no automobiles, no street cars, no railroad trains will be found on the 'Fairy' Isle." It was a place where "nature at her happiest mood" could be enjoyed on "delightful drives through Michigan's 1,000-acre State Park."[88] The island's "storied past" was highlighted. The publication split the difference between the Grand Hotel's public claim to be a watering place for the wealthy and the park commission's hope to supply rest to teachers, shop clerks, and farmers. "Luxurious hotels in varying sizes and styles and moderate-priced boarding houses offer hospitality as varied as the tastes and requirements of the Island's guests."[89]

Agreement on a tourist brochure, however, did not signify agreement on underlying issues. Conflicts remained and the Civic Association was unable to broker a lasting agreement between the island's factions. Within three years of its founding, the association embraced the commission's plan to gain control of the waterfront and remove the buildings between Main Street and the beach.[90] The Civic Association had come to support the park commission's views.

Despite its collapse as a forum for the open exchange of ideas, the Civic Association marked the beginning of a healing process. Future disagreements between the commission and island business owners and residents would never be as bitter as those that took place in the 1920s. Grudgingly or not, the businesses came to understand that as part of the commission's desire for a people's park, the commission would interfere when private businesses, in the commission's opinion, significantly threatened the commissioners view of how the island served the public. The commissioners, in turn, came to realize they would never own the entire island and never hold final authority over private enterprise within the village. They would need to, sometimes grudgingly, take into consideration and accommodate the views of business owners and residents.

If a lasting peace between the commission and the tourist industry was not hammered out by the end of the 1920s, boundaries had been created that defined future conflicts.

And despite the turmoil, through the efforts of the Mackinac Island State Park Commission, what had been a national park was preserved largely intact. Commercialism that might have destroyed the park was kept within reasonable limits. As a result, the state park overseen by the Mackinac Island State Park Commission continued to enjoy the well-deserved reputation for natural beauty that had led to its originally becoming a national park. That park, enhanced by a now well-developed island resort industry, continued to be an anchor for tourism.

CHAPTER 4

New Leaders and New Directions, 1929–1941

The Depression years were difficult on Mackinac Island, reflecting a national crisis within the hospitality industry. Times were hard for the small merchants who ran stores on the town's main street.[1] Otto Lang, who had come to the island in 1922 and remained for more than a half century, reminisced that during the Depression about half the store fronts along the village's main street were vacant.[2] Marvin May, who became one of the "fudge barons," and had come to the island in 1930 when his father became an employee in the Murdick fudge shop, recalled that during the Depression, you could "rent half the town."[3]

Small businesses suffered greatly, but what more often caught people's attention were the plight of hotels. In 1933, the profitability of hotels across America had declined to approximately half of what it had been in 1929.[4] Seasonal hotels, like those on Mackinac Island, saw an even larger decline. Between 1929 and 1933, seasonal hotels experienced a 75 percent decline in receipts.[5] On Mackinac Island, the small Cannonball Inn closed in 1938.[6] The Island House Hotel teetered on the brink of bankruptcy throughout the decade, eventually falling into receivership.[7] But what

Cannon Ball Inn, undated. Detroit Publishing Co. COURTESY THE LIBRARY OF CONGRESS.

Island House Hotel, ca. 1900. COURTESY CLARKE HISTORICAL LIBRARY, CENTRAL MICHIGAN UNIVERSITY.

sent a chill up the spine of everyone on the island was the bankruptcy of the Grand Hotel.

During the Roaring Twenties, illegal gambling and illegal liquor sales made the Grand Hotel profitable. The Depression ended those happy days, and by 1930, the Grand had few guests to spend their money there, legally or illegally. A dismal record was set on Friday, June 16, 1930, when the Grand Hotel's two hundred employees served sixteen guests.[8] In an effort to attract new business, the room rates for the 1931 season were reduced 20 percent.[9] The reduction had no effect on the bottom line. In October 1931, the Grand Hotel's owner placed the institution into voluntary receivership to protect it from legal action brought by those to whom it owed money.[10]

In 1932, the hotel's court appointed receiver, Otto Lang, persuaded E. J. LaChance, a former co-owner of the hotel, to run the Grand for the season. LaChance made a valiant effort to dig the hotel out of its financial hole.[11] The highlight was a marketing campaign arranged between the hotel and the Great Lakes Steamship Company. The steamship line tailored its lake cruises so that each one included a two and a half day stop-over on the island, with passengers staying at the Grand.[12] A lavish advertising campaign appeared in the *Detroit Free Press*. The campaign combined frequent and enthusiastic stories about the island and the hotel, little more than advertising copy masquerading as news, and traditional display ads. In the July and August editions of the paper, some aspect of the campaign appeared almost daily.[13] Hotel officials kept up a good front, announcing in July that the hotel had "a very good guest list,"[14] although they knew the truth was far different. LaChance's daughter, who worked in the hotel that summer, recalled, "It was a grim summer, at best."[15] When the Grand Hotel closed at the end of the 1932 season, it seemed unlikely it would reopen.

In the spring of 1933, the hotel was offered for sale at a court-supervised bankruptcy auction. Only one bid was received, submitted by W. Stewart Woodfill.[16] In later years, Woodfill liked to tell the story about how people thought he was crazy to buy the hotel. Even his wife, Elizabeth, opposed the idea.[17] In truth, Elizabeth, and most other people, were not too far off the mark. The hotel's centennial history described the establishment in 1933 as "debt-ridden and dubiously run."[18] The terms of the purchase were challenging. Woodfill had to pay up front $25,000 in cash. He also was required to pay all back taxes that were owed and assume all the existing hotel debt at face value. Until the taxes and debts were paid in full, dividends on hotel stock were forbidden and employee salaries were strictly regulated.[19] The idea of buying the hotel likely made sense to Woodfill because he was among the hotel's largest creditors.[20] In promising to pay back all outstanding hotel debt, he was promising to pay back himself. But perhaps Woodfill bought the hotel because he was betting that, for all its problems, Mackinac Island remained "Mich[igan]'s show resort,"[21] with the Grand Hotel being the center of the show. He was betting on a bright future for Mackinac Island during a dark time.

It took Woodfill six years to make good on the terms of the purchase.[22] To do so, he employed tactics that ranged from showy to shady to simply illegal.[23] Woodfill's most immediate problem was the people to whom the hotel owed money. He wrote to each of them asking for more time to pay off debts and most agreed to offer him extended payment terms.[24] A few creditors, however, refused Woodfill's request and demanded full payment of the money owed them. How Woodfill dealt with one of those companies demonstrates what a bad idea this was.

Woodfill launched a memorable public tantrum when the electric company was unwilling to postpone his debt. When bluster did not change the demand that he pay in full, Woodfill announced that he would have nothing more to do with the electric company that served the island. Woodfill purchased a diesel generator to make his own electricity. It was a good start, but he needed the electrical transformers servicing the hotel building to get the power from the generator to the hotel rooms. Most likely the transformers were owned by the electric company, but Woodfill created a verbal fog that left the point uncertain. In the end, the electric company, apparently tired of dealing with the man, abandoned the transformers. Woodfill ran the generator for about three years before coming to an understanding with the electrical company and reconnecting his hotel to the company's power lines.[25]

If vendor costs were one of Woodfill's concerns another was the cost of labor. Woodfill had a large payroll but a need to keep employee costs low. To accomplish this, he dealt harshly with his employees. He promised employees a salary of twenty-five dollars a week—but workers actually received eighteen dollars each week. The seven-dollar difference was held by the hotel and paid only if a worker stayed for the entire season.[26] Woodfill made good on the promise to pay the withheld money to those who stayed. But Woodfill knew that most of the many students he hired would need to leave when their fall semester began and could not stay until the end of the season. He pocketed the seven dollars a week difference taken from the wages of those students, and any other worker who left early.

Withholding wages, for one reason or another, was a practice that would remain common for decades on the island. Woodfill, however, went one step further. On occasion, he didn't pay any wages. One summer during the 1930s there were fifty musicians and dancers that never received any salary. "Pay" for them, Woodfill decided, was room, board, and what they could get in tips.[27] It is unlikely they agreed to this, but they had little choice in the matter. As late as the 1950s, if Woodfill was sufficiently angered, he would dismiss an employee or a group of employees on a moment's notice. R. D. "Dan" Musser, Woodfill's nephew and the subsequent owner of the Grand Hotel would recall, "You could fire a guy and you could say, 'OK Mr. Porter, you're done here and your check will be waiting for you at the 3:00 boat.' And you could send the checks down there to the cops and they'd give it to you and make sure you got on the damned boat."[28] Woodfill did what he felt he had to do to control labor costs.

Abusing both creditors and employees, however, did not generate enough money to pay the hotel's obligations. Islanders knew this and rumors again circulated that the Grand Hotel might not reopen.[29] To increase revenue Woodfill continued illegal gambling at the Grand. He had little choice. As part of his purchase of the Grand in 1933, he had made a verbal deal with LaChance and Olive Ballard, who represented the Ballard family, that divided the proceeds from the gambling. For ten years LaChance would receive 10 percent of the gross gambling revenue. During the same decade, Woodfill and the Ballard family would equally divide the remaining gambling money.[30] And there was money to be made. Dan Musser estimated that during the 1930s, in a "good year," Woodfill earned about $20,000 from illegal gambling, approximately $750,000 in 2023 dollars.[31] A bit of arithmetic, based on the 1933 agreement, suggests that total hotel's gambling revenue in a good year was about $1.65 million in 2023 dollars.

Publicly, local officials denied that gambling was ongoing. In 1939, for example, the county prosecutor and the island's mayor jointly announced, "Gambling is practically extinct at Mackinac Island."[32] They knew better. In fact, the community was taxing gambling. The city had enacted an "amusement tax" that was primarily levied on slot machines.[33] The man who was Mackinac County prosecutor from 1933 to 1941, and who included Mackinac Island in his jurisdiction, would reminisce years later that unless some local resident complained, he simply ignored the gambling. "If it would keep the hotel going he was for it."[34] Officials in Lansing also knew what was happening, and generally remained silent, unless they found it a useful political issue. For example, in 1938, Michigan's attorney general obtained a good deal of favorable newspaper coverage when he ordered a statewide crackdown on illegal gambling, specifically mentioning Mackinac Island.[35]

Eventually Woodfill ended organized gambling within the Grand. Perhaps he was never truly comfortable with it or, at least, with the possibility of legal action against him and the hotel. Years after the gambling ended, Woodfill was pressed by a hotel guest to reminisce about it. Woodfill hesitated, then answered, "Such things were forgotten, and nothing could mar the elegance and charm of [the Grand Hotel]."[36]

If both the island's hotels and merchants were struggling, so too were the summer people living in the grand cottages that had been built on the bluffs. As one person put it, the most wealthy residents "seemed to have vanished."[37] The problem became so acute that the park commission drastically reduced rental fees.[38] Before the Depression, the annual rent charged for a lot was between five hundred and six hundred dollars a year. The commission reduced the rent to as little as sixty dollars.[39]

It was not enough. Eventually the park commission seized abandoned cottages for unpaid rent and resold them.[40] Over time the commission found purchasers for the buildings, but the wealthy individuals who had come to the island at the beginning of the century never came back. The new owners of the cottages were also wealthy, but they had no ties to the elite community that had been founded at the beginning of the twentieth century. It was a profound change.

The Mackinac Island State Park Commission was also experiencing profound change. The law that created the commission in 1895 stated commissioners would be appointed for a ten-year term.[41] In 1927, the law was changed. Commissioners became political appointees who served at the pleasure of the governor.[42] The result was a complete loss of membership continuity.

Governor Fred Green, who signed the bill into law, did not use his new authority. The reason may have been a very important gathering about to occur on the island. The legislation became law June 2, 1927. On July 25, 1927, the National Governors' Conference was to convene on Mackinac Island. President Calvin Coolidge had announced plans to attend, as had the two men everyone expected to run for president in 1928, Herbert Hoover and Al Smith. Green may have been concerned that a wholesale change in the commission's membership weeks before the Governors' Conference would cause problems. Whether he was worried about the Governors' Conference or something else, Green kept in office the men already seated on the park commission.[43]

In 1931, newly elected Governor Wilbur Brucker appointed new commissioners. Each of his successors did the same and, as a result, continuity within the commission was lost. The change within the commission was made worse because the 1930s were a particularly volatile moment in state history. During that decade, Michigan's governor was elected every two years. In the 1930s, every two years voters vented their anger and frustration over the political establishment's failure to help them during the Depression by electing a new governor from the party out of power. In 1932, Republican Brucker was replaced by Democrat William Comstock. In 1934, Republican Frank Fitzgerald became governor. In 1936, Democrat Frank Murphy was elected. In 1938, voters returned Republican Frank Fitzgerald to the governor's office.

Each newly elected governor replaced his predecessor's commission appointees with new members. As a result, between 1931 and 1939, twenty-two individuals were appointed to the commission. Of those appointed between 1931 and 1939, only three members served more than two years during the decade. In contrast, between 1895 and 1930, twenty-three individuals served on the commission.[44]

Making matters worse, each new group of commissioners appointed a different park superintendent. Superintendent Frank Kenyon, who had served since 1909, was dismissed in 1931 and replaced by Clarence

LaChance. LaChance was the first of five different individuals who would serve as park superintendent between 1931 and 1939.[45] In 1937, the legislature voted to establish five-year, fixed term appointments for park commissioners.[46] The act, however, had only limited immediate effect. In 1939, all of the commissioners appointed to fixed terms resigned, allowing the new governor to appoint five new individuals as commissioners.[47]

With businesses closing, cottagers leaving the island forever, and the park commission membership changing every two years, continuity with the past was lost. New leaders, with new ideas, emerged. Three men became prominent: W. Stewart Woodfill, Wilfird "Bill" Doyle, and Otto Lang. Each would spend decades in positions of importance on the island.

Woodfill was a person people remembered. A young man who first met him in 1952 wrote, "He was over six feet tall and stood very erect. He was impeccably dressed in a double breasted blazer and coordinated handkerchief, white pants with razor sharp crease, and black and white shoes. He wore dark, horned-rimmed glasses and carried a walking stick.[48] He spoke with an air of great authority."[49] Woodfill was born in Indiana in 1896 and was part of a prosperous family. His grandfather was the past president of the local bank. His father was a prominent local merchant.[50]

Woodfill recalled staying at the Grand in 1907 when he was a boy. His family took him to the hotel hoping to ease his severe seasonal allergies.[51] In 1919, he applied for a summer job at the Grand Hotel. He was hired as a desk clerk where he demonstrated "a flair for the routine."[52] In addition to a flair that charmed customers, he was a hard worker who was interested in learning everything he could. He volunteered to work extra hours in the auditor's office, run by his eventual business partner E. J. LaChance.[53] Woodfill was rapidly promoted and was soon an assistant manager. He unexpectedly became the hotel's general manager in 1921, promoted when the man who had been hired for the position died.[54]

Woodfill learned other skills in addition to entrancing customers and mastering the business side of a hotel. The Ballard family taught him how to make money illegally. Woodfill would say that he "worshipped" the family patriarch and mastermind, Edward Ballard.[55] When James Logan Ballard died in 1923,[56] the family member in whose name the hotel had been placed, Woodfill saw a chance to buy an ownership stake in the Grand. Eventually, Woodfill, E. J. LaChance, and Edward Ballard succeeded James Logan Ballard as the hotel's owners. Woodfill managed the operation. LaChance served as hotel president. Ballard was ostensibly a silent partner, but, in all likelihood, he oversaw the hotel's illegal activities.[57]

In 1927, for reasons he never explained satisfactorily, Woodfill sold his share of the hotel to Ballard and LaChance. He received some cash but most of the money owed him was paid with a promissory note.[58] Woodfill still owned a cottage on the island where he spent the summer, enjoying the pollen-free air and likely keeping an eye on the hotel.[59] In February 1933, Woodfill bought the institution.[60]

For all of his hardheaded economic sense, what Woodfill most needed to do to turn the Grand Hotel around was attract guests. To do that Woodfill relied on his gift for public relations. He explained in a letter to his father, written in 1934, that the hotel's attractiveness to customers was enhanced by "illusions" and the "impressions" left by advertising and rumors.[61] Woodfill was not the first manager of the Grand Hotel to realize the need for good public relations, but he may have been the greatest master of it to ever run the hotel. Woodfill spent his entire career at the Grand creating illusions, impressions, and rumors that surpassed any imagined by his admittedly wily predecessors.

Woodfill skillfully manipulated reporters. In 1935, the Michigan Press Association annual meeting was held at the Grand. Woodfill was happy to host the event which brought approximately two hundred journalists and their families to the hotel. He was likely happier when the group's visit led to identical stories about Mackinac Island. Written by the organization's president, the articles appeared in newspapers around the state, often on the front page.

> Many of our newspaper group came to the Island for the first time, and were deeply impressed with the natural beauty of the Island, situated as it is between Old Mackinac City and St. Ignace.[62]

It was the kind of publicity that Woodfill hoped would help the Grand Hotel and the island's tourist economy survive the Depression.

Manipulation of the press was a skill he possessed, but he also was skillful at creating rumors. In the 1920s, Woodfill "let slip" that the then very popular movie star, Gloria Swanson, would be a guest at the Grand. Gloria Swanson was likely nowhere near the state of Michigan on the day of her supposed visit, but Woodfill appeared on the hotel porch escorting an elegantly dressed, heavily veiled, woman, who looked a lot like Swanson. Woodfill, showing a suddenly developed concern for guest privacy not shared by his predecessors who had routinely published guest lists in Chicago and Detroit newspapers, refused to reveal the lady's name, but he also did not deny that she was Gloria Swanson.[63]

Sometimes Woodfill went beyond spreading rumors and just lied. In 1935, he advertised that "strictly fresh vegetables were served" in the dining room, except for the canned ones needed to "ride out the emergency." In 1938, he advertised that guests could ride north in air-conditioned trains, an impossibility since the railroads did not operate air-conditioned cars north of Chicago or Detroit. If caught in an exaggeration or a lie, Woodfill usually claimed what he had said was a harmless prank, and, in that vein, his greatest prank took place in 1936. Based on Woodfill's private statements, *Ripley's Believe It or Not*, a widely syndicated newspaper feature, shared that the hotel's porch was the world's longest, measuring 880 feet. For years afterword, when asked about the porch, Woodfill pointed the questioner to Ripley's story. However, much like Woodfill knew the veiled lady he escorted on the Grand Hotels porch was not Gloria Swanson, Woodfill almost certainly knew the hotel porch was not 880 feet long. The fabrication was exposed in 1982 when an architectural survey stated that the porch was 628 feet long. When asked about the missing 252 feet, Ripley's editors said that they did not fact check the claims they received.[64] W. Stewart Woodfill apparently was not available for comment, as no newspaper published a word from him on the matter.

Woodfill was a talented businessman, one unafraid of using sharp elbows, shady tactics, or lawbreaking when needed. He was a master publicist who could sweet talk a newspaper reporter into writing a complimentary story or make up a tall tale and find a way to enshrine it in a legitimate source. But the 1930s also revealed he was a man who could get things done. His determination kept the Grand Hotel in business. Woodfill would become an island legend. So too would another man who would emerge in the 1930s, Bill Doyle.

Doyle was from Menominee, Michigan, and was part of a politically well-connected family. He served one term in the state senate in 1933 and 1934.[65] In 1935, he became a lobbyist and spent the remainder of his career demonstrating an uncanny ability to ensure that state laws came out right for the groups who paid his salary.[66] When he retired in 1961, a poll taken among fellow lobbyists and members of the legislature found that they thought him the most effective lobbyist in Lansing. One Detroit newspaper said the same, writing that his retirement came "after 26 years as the most successful persuader the Michigan lawmakers ever met."[67]

Doyle's career was in Lansing, but his love was for Mackinac Island. One of Doyle's earliest visits to the island was a very mixed affair. As a Republican candidate for state senate in 1932, he paid a visit to Mackinac Island, which was part of the district from which he was seeking election. The island generally voted Democratic, but the reception he received was warm. About to board the ferry for the mainland, Doyle was enjoying a chat with the Democratic leader James Doud. Realizing that Doyle may have drawn the wrong conclusions, Doud felt the need to set things right. He put Doyle on the ferry with the words, "But so there'll be no misunderstanding, we're all voting Democratic."[68]

Elected to the state senate despite the Democrats on Mackinac Island, Doyle sought political gain by attacking Democratic Governor William Comstock's newly appointed Democratic park commissioners. Doyle claimed that "intolerable conditions" existed on the island.[69] He alleged that there was "open immorality at some places on the Island" and he demanded an investigation into the commission's finances. With ongoing gambling at the Grand Hotel and elsewhere, there was some truth to Doyle's allegations. The commissioners, however, like so many public leaders had earlier and would later, denied any illegality or wrongdoing. The commission's chair dismissed Doyle's accusations as "purely political,"[70] which also was likely true. In any case, no investigation was held.

Despite the Democratic majority and his mudslinging against the park commissioners, Doyle and his wife fell in love with the island. TDoyle and his wife, Catherine, fell in love with the island. They were not uncritical of its faults. As Catherine would reminisce, when they first came to stay in 1935, they found the island was "a dusty old antique stuck away in the attic." A place, she said, that was "dying on its feet."[71] But it was an antique that appealed to both of them, and they first rented and eventually owned island property.[72] Doyle would spend the rest of his life shaking the dust off the island. In the 1930s, he would become a regular guest at the park commission meetings.[73] In 1939, Governor Luren Dickinson would make him a member of the Mackinac Island State Park Commission.[74] He remained on the commission, more or less continuously, until 1985.

The third man who would help reshape the island was Otto Lang. Lang arrived in 1922, sent by an accounting firm to help untangle the complicated financial situation created by the death of George T. Arnold, the past park commissioner who owned a ferry line and extensive property on the island. The job was supposed to last three or four months. Lang, however, so impressed Arnold's widow that she hired him as her accountant, made him manager of the ferry line, and also made him manager of the many other island properties she owned.[75] When Mrs. Arnold died in 1932, her heirs liquidated the estate. Lang and Prentiss Brown of St. Ignace jointly bought the ferry line as well as most of the property owned by Mrs. Arnold. Lang served as the operating partner of the boat line until 1973.[76]

Lang was a leading member of the island's business community and seemed to have had his fingers in most every civic endeavor within the community. In contrast to the broad dreams and sometimes overreaching aspirations of Woodfill and Doyle, he "sensed accurately what were feasible goals and what were fantasies." He had his critics, people who found him "aloof, shrewd, calculating, and a tight-fisted bargainer," but a newspaper editor who knew him well dismissed this as criticism coming from those who "did not know him well enough to experience his warmth, good sense, and to enjoy his sense of humor."[77]

Beginning in 1934 Woodfill, Doyle, and Lang began to reshape the island. Woodfill, as the very visible owner of the island's largest hotel, often took the lead. In September of 1935, for example, Woodfill spoke at a park commission meeting about a plan for the "improvement of the business conditions on the Island." By improvement Woodfill seems to have meant improved island publicity. Doyle was in attendance at the meeting and he supported Woodfill's ideas, complaining about the lack of printed tourist information.[78] After the meeting, for the first of what would be many times, Doyle took advantage of his presence in Lansing to

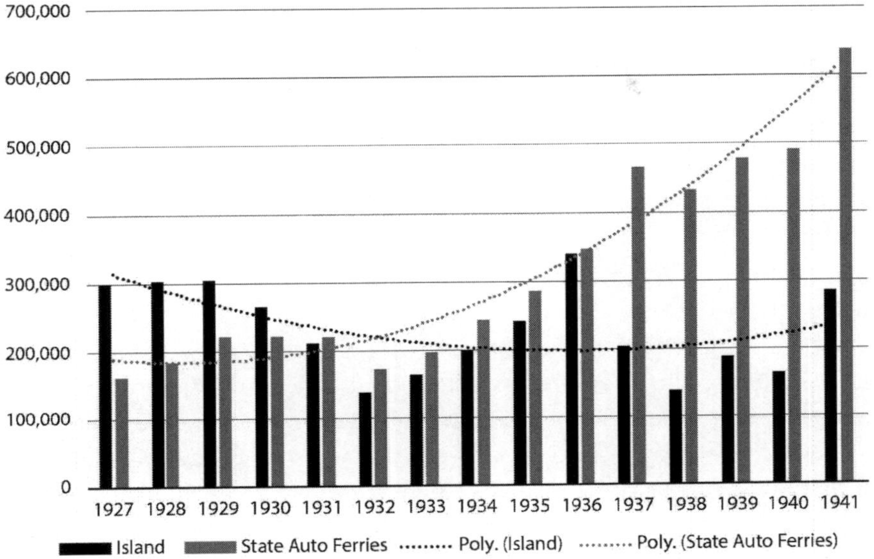

Selected ferry passengers in the straits, 1927 to 1941. The US Army Corps of Engineers recorded the number of passengers visiting Mackinac Island from 1915 until 1989, with the exception of the years 1952 through 1956. The corps also records the number of passengers carried by the state auto ferries across the straits from 1927 until 1951. These statistics can be found in *The Report of the Chief of Engineers, United States Army, Commercial Statistics*, published annually through the year 1952. Thereafter the corps removed the Commercial Statistics from the *Chief Engineers Report* and placed them in a new annual publication, *Waterborne Commerce of the United States*, which first appeared in 1956, but did not begin including passenger traffic to Mackinac Island until 1957. The statistics found in this chart are drawn from these publications.

work on behalf of the island, in this case to acquire an appropriation for publicity.[79] Woodfill and Doyle soon became a team, which Lang became formally affiliated with in early 1936, when the three men became the park commission's advisory committee on "advertising and publicity."[80]

What particularly concerned the three men, and the rest of the island's tourist industry, was not simply that the number of people coming to the island was falling, but this was occurring when the number of people visiting the Straits of Mackinac was increasing.[81] In 1929, the last pre-Depression season, the state's auto ferries carried 222,821 people across the straits, while other ferries brought 306,112 people to the island. In 1941, 637,310 people rode across the straits on the state's auto ferries, while ferries carried 285,633 people to the island. The gross numbers and the long-term trend of fewer people coming to Mackinac Island were frightening to anyone doing business on the island.

Most within the island's business community blamed the decline in business on two things. The first was a public perception that the island was expensive. This concern appeared with unsettling regularity. In 1935, Governor Frank Fitzgerald complained about how expensive it was to live on the island, calling it a rich man's resort.[82] In 1939, Governor Luren Dickinson repeated the same complaint.[83] It was not only two governors who thought the island was too expensive. In August 1939, several Michigan newspapers ran a story written by Gene Alleman, a syndicated columnist, who wrote that people were bypassing the island because they believed it cost too much to visit.[84]

In September 1935, there was discussion at a park commission meeting about the need to "kill the idea" that a vacation on the Island was too expensive.[85] The story that was usually spun by the island's advocates was that a vacation on Mackinac Island was either surprisingly affordable, or at least a good deal for what you got.[86] As for Governor Dickinson's complaint, a letter to the editor was printed in one of the Detroit newspapers shortly after his statement that admitted you could spend a lot of money on the island, but claimed you did not have to. As the writer put it, "Perhaps the Governor stayed at the Grand Hotel which is expensive, but it is not by any means the only place on the Island."[87]

For all the smoke they generated with phrases like "surprisingly affordable" and "a good deal for what you got," business owners knew there was something to the perception. Woodfill and others on the island lowered their prices. In 1940 Woodfill announced that for the first time in its history, the Grand Hotel guests could book a room "European plan." The Grand had always been "American plan," room and meals. But now guests, if they chose, could eat in a low-cost cafeteria Woodfill had opened in the hotel to supplement the hotel's main dining room, which was reserved exclusively for his American plan guests. Guests could also go elsewhere for a meal. Woodfill likely gritted his teeth when he announced the plan, and he gritted them even harder when he shared that about half of the Grand's guests were taking advantage of it.[88] Woodfill's cafeteria signified how island businesses were lowering costs, even if they preferred not to discuss it very much.

Automobiles were the second issue bedeviling the island's tourist trade. As early as 1926, Woodfill, then general manager and part owner of the Grand Hotel, was placing ads advising potential visitors, "Don't miss it [the Grand Hotel] or your motor trip will not be complete."[89] The island business community watched with growing concern the increasing number of tourists coming north in automobiles. These travelers sought a less expensive way to vacation; camping or staying in an overnight cabin. Although no one on the island expressed the problem in terms of class, the nation's working class could increasingly afford to buy inexpensive automobiles. Working class families soon realized that the car could be used to take what for most was their first vacation. These new travelers, however, had even less discretionary income than the already price-sensitive middle class. As a result, working class tourists lowered the price of a vacation by camping or staying in an overnight cabin, low-cost amenities that many in the middle class also increasingly took advantage of.[90]

If the island's tourist industry never discussed the problem in terms of class, they did frequently discuss the problem of automobile tourism, and some merchants began to call for a radical change, allowing cars on the island. Until that happened, they agreed with what had been written in an editorial published in *Hotel Month*, "Until such a service [auto ferries to Mackinac Island] is provided by state, or private interests, many who would otherwise visit the historic and restful Island will pass it by in their eagerness to be 'on the way.'"[91]

The island's ban on automobiles originated at the beginning of the twentieth century.[92] Continuing the ban was an idea that had many friends, particularly among the politically influential carriage operators, who supported it at a meeting in 1929.[93] However, others were of a different opinion and Park Superintendent Frank Kenyon developed what he likely believed was a compromise on the subject. Kenyon proposed that the Michigan state automobile ferries add a stop at Mackinac Island, at an automobile camping area built by the park commission. The cars, however, would not leave the campground. To ensure they stayed there, the campsite was to be surrounded by a fence.[94] Kenyon's proposal created a convenient way for the motoring public to reach the island, but kept their cars safely corralled.[95] His plan was adopted by the park commission, who authorized building a dock at British Landing for the state auto ferries. In the fall of 1928, state ferries began regularly scheduled stops at the new dock. State ferries also ran to the island in the fall of 1929.[96]

The state ferry service was unenthusiastic about the new stop. Adding an island docking delayed its boats in performing their primary function of shuttling cars across the straits. But the real controversy was on the island itself, where many merchants wanted more than just a ferry stop and a car corral. In 1929, Superintendent Kenyon explained the situation this way: "Many of them [business people] are of the opinion that if cars are allowed to run there [on Mackinac Island] things will be much better financially."[97] Newspaper reports in late 1929 claimed that the mayor and city council, worried about the island's future as a tourist destination, had decided to support allowing cars on the island.[98] Feelings ran high, and there was talk of a referendum of voters on the issue.[99]

Whether or not to allow automobiles on Mackinac Island eventually became a statewide discussion. In 1935, the *Clare Sentinel* reported that new efforts would be made to popularize Mackinac Island and the most likely change would be lifting the ban on automobiles. The paper claimed that Governor Frank Fitzgerald "has indicated that he favored lifting the auto ban to some extent at least."[100] In 1935, the *Detroit Free Press*, editorially noted that for the government of the self-proclaimed auto state to ban cars at Mackinac Island's state operated park was "a bit incongruous." But the issue was touchy even in far-away Detroit. The paper's editorial board, who regularly lectured state and local officials on what their course of action should be, simply wrote that the park commission had a tough choice to make.[101]

In 1939, the statewide conversation about the Island's auto ban resumed. The *Saline Observer* wrote,

> In this speed-craze age of automobiles, Mackinac Island is one place in Michigan that has defied mechanical modernization. No automobiles! Such is the restriction imposed by the town council with the moral support of Island summer residents and the state park commission. . . . Harvey

> Campbell, Detroit Board of Commerce, thinks the Island is destined to have a continued decline in tourist trade largely because, as he colorfully phrased it, "the Island rests on a manure-dump." . . . His remedy is the Detroit-made sightseeing bus, used at the New York and San Francisco world's fairs, with fares at popular prices. Sentimental lovers of Mackinac Island rise in wrath at the idea.[102]

The *Saline Observer* summarized why this was a topic of conversation, saying, "Auto tourists, were passing by the Island."[103] The disastrous year of 1938, in which the number of visitors numbered only 760 more than at the nadir of the Depression in 1932,[104] led many on the island to agree. In November 1939, a petition signed by two hundred islanders asked that the state ferries resume service to British Landing. That "the Island hotel business has suffered in the past few years" was proof, according to those who signed the petition, that things had to change.[105]

The park commission offered little help in resolving this issue. It had allowed Kenyon to build the dock at British Landing and land cars there in 1928 and 1929. One of its members had even explored the possibility of profiting from a new carriage route from British Landing to the city.[106] But the chair of the commission, G. A. Hendricks, who had once been a supporter of Kenyon's plan, said in early 1930 that he would "fight to the last ditch to keep [automobiles] off the Island roads and paths."[107] Commissioner Hendricks would explain his position in a letter written to the governor in May 1931: "The quaintness of the place due to the old fashioned carriages and horses has a tourist value not equaled anywhere in the United States."[108] In 1934, the democratically appointed park commissioners voted to not resume the state ferry service at the British Landing dock.[109] Perhaps those in favor of the idea hoped that a new group of commissioners would reverse this decision, but a similar resolution was passed in 1935 by a Republican-run park commission.[110] This bipartisan opposition to automobiles wavered in 1938, when the commission included in a request for federal funds money to repair the dock at British Landing "with the hope that it would become a part of the Straits Highway Ferry service between Mackinaw City and St. Ignace."[111] The feckless commission of the 1930s was a mere shadow of the forceful body that in the 1920s had attempted to own the entire island.

In the end, those who wished to allow automobiles on the island lacked the political strength to change the status quo. This was because they failed to persuade the three men who would emerge as the island's new leaders, Woodfill, Doyle, and Lang, to support the change. Woodfill likely saw the lack of cars as a mark that helped distinguish the island as a high-class resort. Doyle likely found the "antique" nature of the island community a charming reprieve from the busy Lansing streets he frequented. As for Lang, as the man operating the island's largest passenger ferry service, the possibility of direct competition with the state ferries would have been an unsettling thought.

None of them believed allowing automobiles on the island would result in transforming the increasing sleepy vacation community into a dynamic tourist attraction. Simply creating a way for tourists camping in Mackinaw City and St. Ignace to drive around the island was insufficient to accomplish that goal. Something else was called for. That something else, they concluded was rebuilding the community's aging infrastructure and refocusing the public's view of the island through the window of history. History, they came to believe, was something that would appeal to middle and upper class visitors and was something Mackinac Island had in far greater depth than any other northern Michigan resort.

In part, the three leaders drew inspiration from how the federal funds pouring into the island's state park were being spent. To help the nation weather the Depression, Washington created many new, well-funded programs. Those programs gave the park commission unprecedented funding. Before the Depression, the state had given the commission approximately $40,000 to spend each year, an allocation that had declined to $14,500 by 1933. In contrast, in 1933 and 1934, the federal government had given the commission $150,000.[112]

So much federal money was being received that Roger Andrews, a Democratic member of the park commission and longtime island advocate, seriously suggested that the federal government resume control of the park. The idea was not as far-fetched as it might sound. Congress had in 1931 authorized the creation of a new national park on Isle Royale, located in Lake Superior fifty-six miles from Michigan's Keweenaw Peninsula. Andrews argued that if the federal government wanted to create an island park in the Great Lakes, resuming control over the readily accessible park on Mackinac Island was a better use of federal funds than trying to purchase privately held land on remote Isle Royale, which would be accessible to few visitors.[113] The federal government, however, never seriously considered reassuming responsibility for the park on Mackinac Island.

Much of the federal money the park commission received was spent on basic maintenance, but some of it supported new, historically related activities, including restoration work of buildings within Fort Mackinac and a substantial increase in the number of historic exhibits within the fort. Even more money was spent on historically related projects in Mackinaw City, where the commission reconstructed the stockade that had once enclosed Fort Michilimackinac.[114]

The park commission's celebration of the 1934 Nicolet tercentenary moved the island further toward the widescale use of history to attract tourists. Jean Nicolet had been the first European to see the straits, and the commission, to draw visitors to its parks, had organized two attractions tourists literally had to come to the island to see: the Nicolet Tercentennial Fair and the Michigan Historical State Fair. Both opened on July 1, 1934. The Nicolet event was organized as many other short-term festivals had been in the past, ending on July 8.[115] The Historical State Fair, however, was to be a summer long attraction, remaining open until September 10.[116] The Nicolet Tercentennial celebration and the Michigan Historical State Fair were successful at drawing visitors, helping the tourist industry begin to recover.

Doyle, Lang, and Woodfill all enjoyed history. Doyle would later express his deep interest in history and that his "pride and joy" on the island were various historical projects.[117] In an article written shortly after Otto Lang's death, a local newspaper editor wrote, "Otto held a passion for historical resources in the Straits area."[118] As for Woodfill, the Grand Hotel's very existence was built around celebrating the gracious lifestyle of days gone by. Thus, they likely all enjoyed the park commission's 1934 history fairs; more importantly, the success of the fairs in bringing visitors to the island showed them that history not only was interesting but also could become a way to make money.

What could also have fired their imagination was the development of Williamsburg, Virginia, as a historic tourist site. Williamsburg had been the capital of colonial Virginia. In 1924, through a gift from John D. and Abby Aldrich Rockefeller, the community began to rebuild and restore itself to look as it had in the years just prior to the American Revolution. The Rockefellers became enchanted with the project and poured millions of dollars into the work. In 1932, Colonial Williamsburg opened its first public exhibition building: the reconstructed Raleigh Tavern, where patriot members of Virginia's House of Burgesses had met during the Revolutionary era. In 1934, employees working at the reconstructed tavern began to dress in period costume, originally in conjunction with President Franklin D. Roosevelt's visit to the site. Within two months after Roosevelt's visit, every Colonial Williamsburg employee who interacted with the public was wearing eighteenth-century costumes.[119]

As they discussed the future, Woodfill, Doyle, and Lang sought a local means through which to perpetuate and control the kind of money the federal government was spending through the commission. The obvious way to fund their dreams were municipal bonds, but paid for in a unique way. The $225,000 bond issue they proposed was to be paid for by a tax levied on each visitor to the island who arrived aboard a commercial vessel. The passenger ship lines would be required to apply a small surcharge to every ticket and turn the money over to the entity that voters were asked to create, the Park and Harbor Commission, which would spend the money.[120] In return for approving the bond and the commission, voters were promised an extensive renovation of the downtown area. One project was to redevelop the island's main boat dock. Another was to construct a 3,900-foot boardwalk. The bonds would also purchase and redevelop a number of waterfront properties. Part of the municipal bond would also fund a "Williamsburg restoration" of the village including the restoration of Old Mission Church and Biddle House.[121] Many years later Lang would call the project "a grand and imaginative effort."[122] Voters approved the bond.[123] Likely to no one's surprise Lang, Doyle, and Woodfill were named members of the Park and Harbor Commission.[124]

Not only was the community to be restored as Williamsburg had been, a new event was conceived to highlight the change. "Old Mackinac Lives Again" was planned for the 1941 season as a weekend festival that would allow visitors to experience Mackinac Island as it had existed around 1820, with the entire community dressed in period costume. The festival became a statewide project. Michigan's governor served as the honorary chair, as well as a costumed participant. The event's working committee was chaired by the head of the Detroit Tourist and Convention Bureau.[125] All four state regional tourist associations agreed to promote the event through newspaper and radio advertising, and the railroad companies also promoted the festival.[126] It was to be "Michigan's most ambitious summer festival."[127] Indeed, organizers announced their hope that it would become a national event "ranking in importance with New Orleans' Mardi Gras, [and] Pasadena's Tournament of Roses."[128] The tourist industry hoped the first festival would bring fifteen thousand visitors to the island. And so things did not stray too far from the vision of Woodfill, Doyle, and Lang, Woodfill chaired the executive committee, and Doyle served as secretary of the organizing committee.[129]

In July 1941, the event began with costumed voyageurs arriving by canoe with their furs. They landed on the city's beach and marched to John Jacob Astor House, once the home of the American Fur Company, then in the process of being returned from use as a hotel into a historical site celebrating its original purpose.[130] Along the way to Astor House the voyageurs stopped for "some pre-arranged fights with soldiers from the fort" and spontaneous "yoo-hooing" at pretty girls. Throughout the town, residents, as well as many visitors, dressed in period clothes. Michigan's governor, Murray D. Van Wagoner, dressed in clothes styled after those worn by Michigan's 1820 Territorial Governor, Lewis Cass.[131] The festival's finale was a ball, called a "fur trader's fete."[132]

The festival brought statewide attention to the plan Woodfill, Doyle, and Lang had struck up to revive the island's tourist industry. If they had their way, they would rebuild the tourist infrastructure and at the same time focus much of the tourist trade on local history. Although they would never use the term, they had developed a way to preserve the island's tourist trade by becoming what later scholars called "a heritage center."[133] They had envisioned the future, but it was a future that was interrupted by World War II and the loss of focus in the immediate aftermath of the war.

CHAPTER 5

The Leadership Divided, 1941–1947

At the beginning of the 1941 tourist season, W. Stewart Woodfill warned, "If America goes to war this year it will not help our tourist business." He added that "war hysteria and uncertainty" was already hurting the industry.[1] Regarding the latter point, Woodfill was wrong. In 1941, more people visited Mackinac Island than in any year since the beginning of the Depression.[2] But he was right regarding the bigger picture. War would be bad for tourism. Avoiding war, however, became impossible when Pearl Harbor was attacked on December 7, 1941.

The island's business community, which unfailingly spun a good story no matter what the real situation was, spent the first two years of the war saying that business was reasonable. Early in the 1942 season, newspaper stories reported that the hotels were enjoying a brisk business, with the Grand Hotel having almost double the number of reservations compared to the same time in 1941.[3] This positive spin continued into 1943 when Woodfill claimed the Grand Hotel was experiencing the heaviest demand for reservations in fifteen years.[4] In August 1943, a Detroit paper printed a story that Mackinac Island was "doing very nearly the same amount of business as last year."[5]

None of this was true. In 1943, the number of visitors to the island was 100,866, the smallest in nineteen years.[6] Close readers of the 1943 newspapers saw numerous clues suggesting business on Mackinac Island was anything but usual. In the spring and summer, newspaper stories ran that the Pennsylvania Railroad and the Pere Marquette Railroad had canceled their special summer tourist season train schedule and that the ferry service from Mackinaw City to the island was "curtailed."[7] The papers also reported that overnight steamship service to the island was reduced.[8]

Trains and boats, however, were not the way people increasingly came north for a vacation. Most vacationers drove their car, and government rationing of gasoline made a family vacation by car next to impossible.[9] But saying gasoline was hard to find did not mean recreational travel stopped. Many Americans still tried to take a vacation trip. And the tourist industry did its best to help them by offering an ingenious way to justify giving extra gasoline to tourists. At the 1943 West Michigan Tourist and Resort Association conference, Woodfill claimed that Michigan's resort centers were performing a vital war service, justifying giving tourists more gas. Woodfill said that the playboys, idle rich and social drones who once vacationed in Michigan were now absent. In their place, he asserted, came sober, tired Americans who found an emotional and physical safety valve in the rest and natural health benefits found at Michigan's resorts. Vacations, in his view, were "vital to victory" and thus worth the needed gas.[10]

It is likely that Woodfill learned this argument from Bill Doyle, who, a few months before Woodfill spoke to the West Michigan Tourist and Resort Association, went with a Michigan delegation to Washington that was led by Governor Harry Kelly. The purpose of the trip was to speak with Office of Price Administrations officials, the people who controlled gasoline rationing, and argue one of the best ways to ensure high productivity among war workers was to give workers a vacation. The Office of Price Administrations officials to whom Kelly spoke stated the obvious: The national fuel supply was fixed, and meeting all of the military's needs required strict rationing among civilians.[11] The tourist industry on Mackinac Island, and for that matter throughout America, would simply have to wait for the war to end before gasoline for vacations again became available.

Gas for vacations was actually Doyle's second idea to bring tourists back to Mackinac. Earlier in 1943, Doyle proposed that the state auto ferries, which ran between St. Ignace and Mackinaw City, be given an additional route between Detroit and Chicago.[12] Claiming the purpose of the service was that it would help ease wartime transportation difficulties, Doyle added that the ferries could make stops "at other Michigan Ports, en route." Left unspoken but clearly implied was that one of those stops would be in the Straits of Mackinac area, bringing tourists from one city or the other to the island. This idea, like the idea of giving workers gas for vacations, went nowhere.

If tourists did reach the island, their stay was nothing like the prewar experience. At the Grand Hotel, the dining room was closed. All the guests ate in the cafeteria that the hotel had opened in 1940.[13] Strolling down the street visitors soon learned that fudge was also in short supply. Wartime sugar rationing limited the amount of fudge that could be made. To stretch the fudge supply, customers were allowed to purchase only one pound of fudge. Limiting sales, however, was not enough to keep the shop open all day. Eventually one fudge maker adopted a war time schedule that saw him making fudge in the morning and selling it in the afternoon. The store would open around 1:00 p.m. and stay open until the fudge was gone, usually around 4:30 p.m. But if a cruise boat pulled into the harbor, the fudge would be gone in an hour.[14] Like tourist destinations across America, the island simply waited out the war, with businesses hanging on as best they could.

With the war's end in 1945, business boomed. The boom had many causes. Wartime production meant virtually everyone who wanted a job had one. Many people, such as women who had never worked before, found themselves employed in order to make it possible to meet very high wartime production goals. But most of the money they made was not spent. Wartime production needs left little capacity for the manufacture of new consumer goods to purchase. Wartime rationing of most goods similarly limited spending. Much of the money being earned ended up in a bank.

Postwar tourism benefited not only from large saving accounts but also from the introduction of paid vacations for most factory workers. Between 1930 and 1940, the majority of America's industrial workers had been granted a paid vacation. Gasoline rationing had given factory workers little opportunity to use their newly granted employee benefit for a vacation, but at war's end, they were ready to get in the car and travel.[15] As one author put it, "With the three things that had become symbols of the good life after the war—a job, a car, and a vacation—Americans hit the road in record numbers."[16]

So many people were going north during the summer that even the railroads, which had seen a steady decline in passengers, tried to capitalize on the "Up North" tourist trade. In 1947, the Michigan Central Railroad began to run a new tourist "special" between Detroit and Mackinaw City. The Timberliner left Detroit Friday shortly after 5:00 p.m. and began its return trip from Mackinaw City late Sunday afternoon. Playing on the "Up North" theme, the train's bar car was outfitted with a knotty pine interior.[17] Many of these vacationers went to the straits region. Mackinac Island experienced a record number of visitors in 1947.[18] In 1948, the Grand set a record for the number of conventions held in the hotel.[19] With money flowing into the cash registers of island merchants, many of the controversies that had roiled the water for years, such as allowing automobiles to drive on the island, were no longer topics of conversation. Good times smoothed over old conflicts.

Perhaps because business was swiftly growing, the coalition of island leaders that had emerged in the 1930s was falling apart. Unity was no longer essential for survival, and strong-willed individuals soon clashed. The first disagreement between Bill Doyle, Otto Lang, and Woodfill occurred in the first years of World War II over the idea of federal contracts to substitute for the income lost by the wartime decline in tourism. Doyle and Woodfill both hoped to obtain federal funds.[20] Doyle suggested that the island would serve admirably as a convalescent center for wounded veterans. Failing that, it would serve equally well as a prisoner of war camp.[21] Eventually something a good deal smaller, an Army weather school, took up residence in the Grand Hotel.[22] Lang, however, did not share Woodfill and Doyle's enthusiasm over significant government activity on the island. He would reminisce years later that "fortunately for the Island, the Armed Services did not take over."[23] When all was said and done, however, the question of military contracts on Mackinac Island during World War II was a minor dispute.

What drove the first significant wedge between Lang and the two other men was the collapse of the boat ticket tax scheme created to pay for the bonds issued by the Park and Harbor commission. Lang had never been completely happy with the tax, recalling that he had preferred the town's residents assume some share of the costs. That idea, however, would likely meet significant opposition among the island's voters. As Lang pointed out, the voters not wanting a tax increase made it "much easier to pass the buck to the boat lines collectively."

Despite his reluctance, Lang agreed with Doyle and Woodfill that the island's economy was in desperate need of stimulus. Whether or not it was the best idea, a boat tax, what became publicly known as the "Doyle plan," was the only practical way to address the problem.[24]

To further address Lang's concerns, Woodfill and Doyle placed among the projects the tax would fund a large carrot for Lang's ferry business. Among the major projects were significant improvements to the Arnold dock. The dock was in need of significant repair. Lang's company apparently lacked the cash to do the work, both because of the very modest amount of ferry traffic during the Depression and because he and his partner needed to devote a significant amount of the income they received to make payments to the Arnold heirs, required annually until 1944.[25] The promise to rebuild the Arnold dock probably went a long way in helping Lang accept the tax used to fund the bonds.

The dock was repaired, but collecting the money to pay back the bonds proved difficult. Out-of-state boat firms quickly and successfully challenged the tax in federal court.[26] With no revenue coming in, the bonds went into default.[27] Lang seems to have played the key role in developing the formula that eventually paid off the bonds. Lang owned the only dock on the island that could accommodate the large ships that visited. He replaced the tax a federal judge had ruled illegal with a perfectly legal, and totally unavoidable, "wharfage fee" that more or less equaled the amount of money out-of-state boat owners would have paid in taxes. Lang also seems to have eventually made the in-state ferry lines running from St. Ignace and Mackinaw City to the island pay, although it was only about half of what they would have owed if the tax was fully implemented. The Arnold line paid its full share plus what the other local ferry operators managed to avoid paying.[28] One way or another, Lang funded the bond, but what it took to do so not only cost him more money than he had expected to pay but also almost certainly left him with much less enthusiasm for any future partnerships with Woodfill and Doyle who could, and apparently did, leave a very difficult problem in Lang's lap.

If Lang was privately unhappy about the bond situation, he was characteristically quiet in public. In contrast, Doyle's behavior as a park commissioner created a bitter divide between himself and Woodfill. In 1939, Governor Luren Dickinson appointed Doyle to the Mackinac Island State Park Commission.[29] As a lobbyist, Doyle had used a quiet conversation in the hall and a well-chosen phrase inserted into a bill pending before the legislature to consistently protect his client's interests. Dickinson likely thought Doyle would have the same subtle touch as a commissioner. Dickinson could not have been more mistaken. Doyle quickly showed that when given official authority he liked to use it—publicly, loudly, and without regard to what others thought. Simply put, he liked to be king, or at least have people think of him that way. The result was that he burned bridges with his fellow commissioners and the governors of Michigan. He also outraged Woodfill, who resisted Doyle's efforts to seize power, leading to what one newspaper called "Lilliputin Wars about who will run Mackinac Island."[30]

Doyle's conduct as a commissioner was routinely outrageous. He developed an ongoing feud with the commission's then chair, Gerald Cleary, that became so intense a newspaper reporter claimed "the two men have made their political enmity practically a hobby."[31] The feud was so public, and so clearly of Doyle's making, that the governor made clear that he would not reappoint Doyle to the body when Doyle's fixed term on the commission expired in 1942.

A lesser person would have been put in his place. Doyle was not a lesser person. Instead, he used his lobbying skills to change the law regarding commission membership. He did this by having passed an innocent sounding amendment to the commission's membership that added to the body a "resident" commissioner, a person who lived on the island. The act also specified that the governor, who officially nominated all Mackinac Island State Park commissioners, was, in the case of the resident commissioner, required to put forward the name of a person selected by island's mayor.[32]

All this might have seemed logical, even benign, to the members of the legislature, to whom Doyle likely spun a tale regarding a small addition to the commission's membership to ensure that the local community had a voice of their choosing at the table. And while the tale was true insofar as it went, it hid a calculated plan. Doyle owned a summer cottage on Mackinac Island. In 1940, he had quietly declared it his permanent residence. Doyle had also become a very good friend of the mayor, who returned Doyle's friendship by informing the governor that Doyle was his choice to serve as resident commissioner.[33] The governor and his fellow commissioners had been outmaneuvered. Doyle still had his seat at the table.

The outmaneuvered commissioners came up with their own plan to silence Doyle. To virtually everyone's amazement, in late 1942, they selected him as chair of the park commission. The "honor" was an effort to muzzle him. The commission's rules stated that the chair was to refrain from debate, unless asked to speak by one of the other commission members. If Doyle played by the rules he would be forced to sit quietly at meetings whenever any matter of substance was discussed, since none of his fellow commissioners were likely to ask for his opinion about anything.[34]

Doyle, however, was a man who read rules carefully, and he quickly found a way around his fellow commissioners' intentions. The commission's rules also stated that the chair was solely responsible for setting the date and time of the commission's meetings. Doyle determined that one or two brief meetings each year were quite sufficient for the commission's needs.[35] In between meetings, Chairperson Doyle personally handled any and all matters of commission business. He established, for all practical purposes, a one-man rule.

As for the inconvenience that the commission was required to submit annual reports to the governor, Doyle discovered there was no requirement regarding the report's content or length. In 1943 Doyle submitted a two-paragraph report to the governor. When asked about its brevity, he responded, "I doubt if any governor has ever taken the time to read any of the reports which have been submitted in the past." In 1944, Doyle again lessened the governor's workload by submitting a still shorter report.[36] Doyle had become, in the words of one newspaper, "a law unto himself."[37] Doyle's one-man rule not only angered the governor and his fellow commissioners, it delayed the development of a group of experienced commissioners who could collectively reverse the chaos caused when partisan politics resulted in commission membership completely changing every two years.

And Doyle, for all his heavy-handedness, knew how to win friends when he wanted them. Governors had been coming to the island for many years, usually staying at a small structure made available by the park commission. In 1944, the year Doyle submitted a one-page commission report to the governor, Doyle also persuaded the legislature to buy a three-story mansion to be operated by the commission and used exclusively by the governor.[38] It would be an investment that would pay dividends Doyle likely never imagined. The Governor's Residence on Mackinac Island became a place several governors grew to love, along with the island itself. Two governors would buy or build their own cottages when their term in office ended.[39] Doyle could not have foreseen this; nevertheless the move was classic Doyle—spite the governor's authority while offering the olive branch of a handsome residence set on a beautiful island.

Doyle made one other important decision, appointing Carl Nordberg park superintendent in 1945.[40] For all his desire to be king, Doyle knew every king needed a strong right-hand man to attend to the day-to-day details. Nordberg proved to be that man. He would serve as superintendent from 1945 until 1966. Although his final years would be difficult, he demonstrated considerable skill in both operating the park and navigating what was often a contentious commission.

While Doyle spent the war years sowing chaos among the park commissioners and publicly thumbing his nose at the governor, Woodfill remained silent about the commission's problems. He was thinking about a future that did not involve the Grand Hotel or Mackinac Island. He made serious efforts to bring together investors to buy two newspapers, first the *Indianapolis Star* and then the *Chicago Daily News*. Both efforts failed, but that Woodfill made them suggests that privately he was less certain of the fate of the island or its flagship hotel than his always bullish public statements suggested.[41] In the end, Woodfill stayed on the island and at the Grand Hotel, but for at least part of the war years, his attention was elsewhere.

When he turned his attention back to Mackinac Island, Woodfill likely realized that Doyle had come into his own as chair of the commission and he would not willingly take a back seat to anyone else. For a time Woodfill, also fond of being perceived as the island's leading citizen, could ignore the issue. That opportunity was given him because of two highly publicized gatherings at the Grand Hotel. In 1943, Republican leaders from throughout United States gathered at the Grand to discuss postwar America. Twenty-two of the then twenty-four Republican governors attended, and most of the party's congressional leadership were there, as were all the leading contenders for the 1944 Republican presidential nomination. The Mackinac Island Conference made history for many reasons, but insofar as the island's tourist industry was concerned, it was both a one-time bonanza of visitors, since the politicians not only brought their aides but were followed by a large press contingent and a huge publicity bonanza. Each day of the conference, and for many months thereafter, the "Mackinac Conference" made national news that frequently mentioned the Grand Hotel but said not a word about the park commission or its chair.[42]

In 1945, Woodfill again basked in the presence of America's elected leadership. The National Governor's conference came to the island during the Fourth of July weekend, bringing the governors of thirty-nine states and two territories to the Grand Hotel.[43] Speculation about whether President Harry S. Truman would join the meeting was constantly in the papers. In May, Michigan's governor announced that Truman was expected to make every effort to come.[44] In June, Truman announced his plans to attend the meeting.[45] However, on July 1, the president sent his regrets.[46] Although it was unfortunate that the event was not highlighted by the first visit to the island of a sitting president, the conference was still a huge publicity success. The participants enjoyed themselves, in part because the Liquor Control Commission "waived the rationing book" regarding alcohol. Even the reporters did well, with one paper noting the "scribes" received several cases of Canadian Club whiskey, compliments of the Distilled Spirits Institute.[47] And again, all these influential visitors were the guests of Woodfill. None of them needed or bothered to consult with the park commission or Chairperson Doyle.

Doyle's growing pique at taking second place to Woodfill would boil over publicly in 1946. In that year, Metro-Goldwyn-Mayer (MGM) Studios picked Mackinac Island and the Grand Hotel as the setting for its "lavish" new film, *This Time for Keeps*.[48] The movie makers brought seventy-five people to the island, including stars and film crew, who it was reported collectively cost $47,000 a day to host on the island. Extras were hired at a rate of ten dollars a day, a good way for a local to make an easy buck. And tourists came in large numbers to see both the stars and the movie making. As the *Detroit Free Press* put it, "Downtown merchants and inn keepers joyously eye the non Islanders who flock ashore to witness a movie in the making."[49]

The film was so good for business that the local authorities waved the rule about motorized vehicles. A sound truck was needed by the moviemakers, and the vehicle was so heavy no group of horses could pull it up the island's steep hills. Eventually, the city allowed the truck to move under its own power at a speed of no more than ten miles an hour. The carriage operators did not much like the exception, and they, or, more properly, their horses, would eventually get their revenge. The noise the horses made caused constant problems when outdoor shooting occurred. It took eighteen takes to shoot a scene of a main character arriving at the island dock. A scene shot on a downtown street took twenty takes.[50]

The problems caused by the sounds of horses, however, were nothing compared to the wrath of Bill Doyle, who seemed to grow more angry by the day watching the MGM film crew take instruction from Woodfill while paying little or no attention to the park commission, or more importantly its chair, Bill Doyle. Late in the evening of July 13 or early in the morning of July 14, Doyle's anger burst out when he berated a senior MGM official. Coming upon each other on the grounds of the Grand Hotel, Doyle informed the official that he didn't care if Michigan's governor or the owner of the Grand Hotel had given the company permission to film. That was not enough, and the movie official "better get over to the fort first thing in the morning and see where they stood."[51]

Instead of getting over to the fort, first thing in the morning, the MGM official contacted the governor's office. The governor, in turn, asked Woodfill to resolve the matter. Woodfill was happy to do so. He and Doyle had previously engaged in a loud, public argument over street closings for the film, one that resulted in "both calling each other beautiful names."[52] Using the authority granted him by the governor, Woodfill wrote to Doyle:

> The Governor advises me that you told Mr. Dietz the picture people would not be permitted to make the picture here unless your approval was secured.... This is to advise you that MGM does not need your approval to make a picture as it can well be made on the hotel grounds, at my house, and in the village, over which areas you and your state park commission have no authority.[53]

Just as during the Mackinac Conference of 1943 and the National Governors' Conference of 1945, Woodfill was in charge. Doyle had "no authority."

Woodfill picked up his pen two days later and wrote a second letter to Doyle, choosing words that likely represented an intentional effort to humiliate Doyle. Reflecting on the fact that the confrontation had occurred at the Grand Hotel, Woodfill banned Doyle from the hotel and its grounds for so long as he was a park commissioner. Not content with merely banning the man from his property, Woodfill added, "I intend to make it my business to see that upon expiration of your term of office as Commissioner, you are never again appointed on the Commission." He did allow, however, that "if you resign from the commission and apologize for your bad behavior you can come to the Grand." To ensure that Doyle understood the matter was personal, Woodfill added, "Be assured that Mrs. Doyle and the children are most welcome anytime."[54]

The whole episode left Doyle humiliated,[55] but Doyle would have his revenge a month later. During the filming of *This Time for Keeps*, all the actors and most of the MGM film crew had stayed at the Grand Hotel. Although unsold rooms remained available at the Grand, to Woodfill's dismay many people, rather than book a room, would visit at the hotel's bar, buy a drink and nurse it for a good long while, hoping to see a star on the cheap. Woodfill put a stop to this practice by announcing that the bar was only open to Grand Hotel overnight guests.[56] Like any regulation, this rule had unintended consequences. In Woodfill's case, those consequences occurred during the annual running of the Chicago to Mackinac yacht race.

Yacht racing was a regular and lucrative part of the island's tourist business. The first major race, from Chicago to Mackinac Island, occurred in 1896 with five yachts competing along a 330 mile course.[57] A second major yacht race, from Port Huron to Mackinac Island along a 242 mile course, began in 1925 with eleven entries.[58] By the 1940s, both had grown in size and the yacht sailors and their many fans made a significant contribution to the summer tourist trade.[59] However, the sailors, when they reached port, were sometimes a rowdy group. A legendary escapade that took place at the end of the 1908 Chicago to Mackinac race was a tale which Woodfill surely knew.

Captain Tom Wilson had won the 1908 race. After anchoring in the harbor, he and his crew used the ship's dinghy to travel the remaining few hundred feet to the beach, where they would begin celebrating their victory. As the tale would come to be told, "The nearest place of interest to the landing dock in the ancient city is the Chippewa Hotel," where captain and crew immediately "betook themselves." "The wide, cool porch" made a fine place to recount their adventures, "and the silent white-clad waiters are most obliging." With a banquet planned that evening at the Grand Hotel for all the yachtsmen, the captain and his crew realized "there [were] many places of interest between the Chippewa and the Grand Hotel . . . and each was thoroughly inspected and enjoyed." Eventually, the by now well-lubricated crew of the winning yacht reached the Grand Hotel where the banquet was accompanied by wine "properly chilled and plentiful."

As the winning skipper of the race, Captain Tom understandably prolonged his stay at the Grand into the "wee small hours" of the morning. Finally, he hired a cab to take him back to the harbor and his boat. Upon reaching the harbor, the captain and the cab driver became involved in a heated dispute over the fare. Captain Tom found the fare asked "extortionate." The captain, "protesting feelingly" to the cab driver, emphasized his point by throwing the driver into the water. Eventually, a fare the captain found agreeable was settled upon and paid.

The captain became even more unhappy when he realized his crew had not only already retired to the ship for the evening, they had taken the ship's dinghy with them. The captain shouted to gain his crew's attention, but they, having endured "a hard race" and enjoyed "a full evening" did not reply. Nor, to the captain's dismay, did anyone else appear to help. The captain's first thought was to "borrow" another dinghy, but he found all the boats were securely locked.

Looking about, the captain saw a boat at a nearby cottage that was dug into the lawn and filled with dirt and geraniums. Filled with "righteous indignation" that any vessel should be so abused, he ripped the boat from the lawn and dumped out the flowers and dirt. All this was "not accomplished with some slight noise" and an occupant of the cottage stuck his head out of a second-floor window. The two gentlemen entered into a conversation, using words that were "spirited, interesting and instructive, pro and con as to the attainments, proclivities and general characteristics of the respective parties."

Whatever the exact words, the conversation did not stop the captain from his task and eventually he hauled the boat to the water. By now the man in the window was more amused than angry, and shouted, "Now that you have got the rotten old tub in the water what are you going to use for oars?" Skipper Tom considered this, and responded, "I guess I'll use your grape arbor for oars." He ripped two long boards from the cottage's grape arbor, promising to return the vessel and oars "in the morning and much obliged to you." The man in the window could only advise, "You better take the pail off the back porch, for I reckon you will have to bail out most of Mackinac Straits through that boat." As the teller of the tale recalled, "It was a battle all the way and just as Skipper Tom reached for the railing of the gangway [of his ship] the flower bed dinghy gave a last gurgle and sank to the bottom of Mackinac Harbor"

Upon subsequent and far more sober reflection, Captain Tom realized that he had overstepped the behavior expected of a captain and a gentleman. He saw to it that all the damage done to the cottage was repaired, and a new flower dinghy, filled with geraniums, put in place. To ensure that the cottage owner never again had a problem with his flower boat, Captain Tom added to his restitution a five-hundred-pound block of granite, to which the boat was securely attached. On the granite was a plaque reading, "With the compliments of Skipper Tom Wilson to the man with the blessed sense of humor."[60]

Captain Wilson's tale, and others like it, should have led Woodfill to realize that if he tried to ban sailors who did not have overnight accommodations in the Grand from the hotel's bar he would be courting trouble. Trouble came his way when several yachtsmen who did not have a room reservation loudly objected when they were refused service. Moving on from the bar, the yachtsmen stood on dining room tables to voice their complaints. They then departed the dining room for the hotel's Terrace Room, where a dance was underway. They stopped the dance by taking over the stage from the band. Woodfill would later assert that many of the yachtsmen were already drunk when they arrived and thus he was within his rights to deny them service, a claim which based on past experience had a good chance of being true, but Woodfill failed to mention was that all the sailors, including the sober ones, were denied service.[61]

Doyle, still smarting from the MGM affair, began what newspapers would call "The Battle of the Grand Hotel Bar." Doyle demanded Woodfill rescind his blanket refusal to serve nonguests. Woodfill refused. Doyle's response was to advise some of those denied service the night of the yacht race to take their complaint to the state liquor commission. Publicly, Doyle needled Woodfill. Playing against Woodfill's always prim image, Doyle allowed, "Snootiness was out of place in the keen, invigorating air of the hotel's cocktail lounge." The bar was open, Doyle argued, to anyone legally entitled to buy "a quick snort." The idea of someone indulging in "a quick snort" in the Grand Hotel likely annoyed Woodfill, but he was more concerned when the State Liquor Control Commission reminded him that that if he wanted to keep the Grand Hotel's liquor license, the bar had to be open to anyone seeking to buy a drink, quick or slow.[62] One suspects Doyle soon also reminded Woodfill that anyone included him, should he happen to accompany his "most welcome" wife and family on a visit to the Grand.

The tempest Woodfill began with the yachtsmen eventually blew over. This was in part because Woodfill realized he had picked a fight with a lucrative clientele and went out of his way to make amends. When *Yachting and Aviation* magazine editorialized that yachtsmen really shouldn't complain but feel complimented since "the Grand Hotel has in late years been interested mostly in the new rich and the suckers who will patronize the gambling casino," Woodfill contritely wrote a letter, published in the same magazine, in which he said that he liked yachtsmen and, as any hotel proprietor must, took action against the few sailors who permitted "their hilarity to run to the extremes of indecency."[63] Although there was some talk among the yacht owners of finding a new finish line for the race, there wasn't a good alternative.[64]

If Woodfill made his peace with the yachting community, he did not do so with Doyle. Woodfill made good on his threat to oust Doyle from the commission. In the island's 1947 municipal election, Doyle's reappointment as resident commissioner became a campaign issue. Mayor Allan Sawyer publicly announced that if he were reelected, he would not reappoint Doyle to the position. Doyle, in turn, claimed Sawyer was "a tool of W.S. Woodfill" who was "sore because I forced him to open his exclusive bar to everybody." Doyle added, likely to save face, that none of this really mattered, since he had already told Sawyer he no longer wanted to be resident commissioner. He may have publicly asserted he no longer wanted the job, but couldn't resist a last swipe at Woodfill, grandly claiming that during his time on the commission, "I was interested in making Mackinac Island State Park a resort for all of the tourists in Michigan instead of a playground for the paying guests of hotels."[65] Shortly thereafter, James Cable was named the new resident commissioner.[66] As for Woodfill, he was "really pleased" by the outcome.[67]

The MGM exchange, the battle of the Grand Hotel bar, and the municipal election of 1947 all revealed how deep the gap between Doyle and Woodfill had become. In the short run, Woodfill was successful in having Doyle removed from the park commission. Doyle, however, was nothing if not a person who played the long game. In 1948, Governor Kim Sigler, without reference to the events of 1947, or that in 1946 he had publicly stated that he would not appoint Doyle to the park commission, reappointed Doyle to the park commission.[68] Why Sigler did so is unclear. Possibly it was because Doyle, who had been outmaneuvered in the 1947 municipal election, was now "moving in on Island politics," leading five members of the city council to write to Sigler saying "the Island needs him."[69] Possibly Doyle offered the governor help in passing bills the governor wanted enacted but which were stuck in the legislature.[70] Possibly it had nothing to do with the island itself or pending legislation but rather was because Sigler, who planned to run for reelection in 1948, needed to restore harmony in the Republican Party, particularly a rift caused by a 1946 law requiring, for the first time, lobbyists like Doyle to publicly register and state who they represented.[71] Whatever was his reason, Sigler gave Doyle what he very much wanted: reappointment to the park commission.

Much of the island's business community likely chuckled in amusement over Woodfill's and Doyle's dispute, and they likely were equally amused by the circus-like atmosphere Doyle had created within the park commission. They could afford to chuckle since business was setting new records on the island. This was partly because of innovative new ideas. For example, Nathan Shayne, owner of the Chippewa Hotel, was exploring the possibility of bringing bus tours to the island. He was acquainted with the "top men" of the Greyhound Bus Company and he and his partner

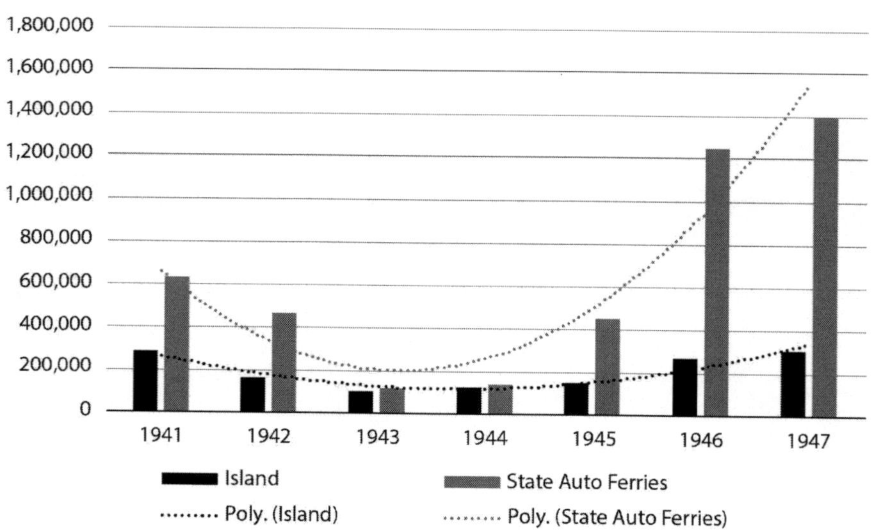

Selected ferry passengers in the straits, 1941 to 1947. Source: *Commercial Statistics: Waterborne Commerce of the United States for the Calendar Year 1944* (Government Printing Office, 1946), 974, 977. *Commercial Statistics Waterborne Commerce of the United States for the Calendar Year 1952* (US Printing Office, 1954), 1188, 1191.

in the tour bus initiative, Otto Lang, worked aggressively to persuade Greyhound to run packaged bus tours to the island.[72]

New ideas like those offered by Shayne increased tourism, but likely more important was the postwar economic boom that was sweeping America. Whatever the reason, after World War II ended, island tourism grew for the first time in fifteen years. But the troublesome prewar trend of more people coming to the straits but, measured by percentage, fewer and fewer coming to the island, continued.

In 1941, about four people visited the island for every ten that took the car ferry between the two peninsulas. In 1947, only about two people visited for every ten that rode the ferries across the straits. Nevertheless, island tourism. Island tourism was again growing. In 1941, 285,633 people had taken a ferry to Mackinac Island. In 1947, 311,325 people had taken the same ferry ride. That was good enough for the island's merchants .[73]

With cash registers ringing merrily, concerns that had dominated the island for more than a decade disappeared. No one was talking about allowing cars on the island. Equally important, the talk about history as a tourist attraction also ended. The idea of resuming the 1941 "Old Mackinac Lives Again" festival seems never to have been discussed. Indeed, talk about history had disappeared early during the war years, as demonstrated by the loss of the island's historic cannons. During a scrap metal drive held on the island in 1942 to collect material for the war effort, the cannons in the fort were donated and hauled away. Exactly who approved the transfer was never determined, but not only were twelve historic cannons that had long resided on the island been given away, three other historic cannons, which had been sent from Lansing to the island for safekeeping, were also turned into scrap. None of the cannons were recovered.[74] History, it seemed, has slipped off the island tourist industry's to-do list.

As the 1940s ended, it appeared that the Mackinac Island tourist trade was returning to a well-tested pattern. With tourism growing, there was little incentive for the business community to seek a new direction. The future, they hoped, would simply be more of the same, with ever-increasing crowds creating ever increased profits. Doyle's one-year absence from the commission and his loss of the chair allowed other members of the commission to begin to gain needed expertise. Carl Nordberg's 1945 appointment as park superintendent began a new era of stability within the park staff. What island tourism most resembled in 1947 looked surprisingly like what had existed during the Roaring Twenties. What would change the picture was the election of G. Mennen Williams as governor.

CHAPTER 6

The Governor Charts a New Course, 1948–1961

G. Mennen Williams, known as "Soapy" because he was the heir to the Mennen Electric Shave fortune, was a phenomenon within Michigan politics. Although born wealthy, he was a progressive politician described by the conservative *Chicago Tribune* as a political reformer who "helped forge the alliance between Democrats, Blacks and union voters in the late 1940s that began a strong liberal tradition in Michigan."[1] His election in 1948 was an upset, attributable to his personal popularity. He became one of Michigan's longest serving governors, remaining in office until 1960. His popularity, however, did not translate into Democratic majorities in Michigan's legislature. As a result, the politics in Lansing were fierce. When Williams succeeded in achieving his legislative goals, it was often by a hairsbreadth.

As part of his 1948 campaign, Williams promised to build a bridge connecting the state's two peninsulas. He called it "Michigan's dream." His Republican opponent called it "Soapy's folly."[2] Whatever it was called, a bridge connecting Michigan's two peninsulas was a more than a decade-old Democratic campaign promise, which had been met with a more than a decade of Republican opposition. Williams tried to make good on his promise about a bridge at the Straits of Mackinac, something that was widely believed to be good for Michigan's Upper Peninsula specifically and the Up North tourist trade generally, but it was a difficult idea to sell. One particularly effective argument that Republicans made that weakened support for the project in strongly Democratic areas like Detroit was by suggesting that if the state was going to spend all the money needed to build a bridge between Michigan's two peninsulas, it would be better spent improving transportation in major population centers.[3] Reflecting this attitude, the *Detroit Free Press* wrote that, "Without getting into the merits of that project [the proposed Mackinac Bridge], we think far more Michigan people would benefit from early completion of the Detroit expressway system."[4] It was a potent argument, one Democrats tried to refute by claiming that once a bridge was built it would save the state money, since the Straits of Mackinac auto ferry system required an approximately $500,000 annual subsidy while the bridge's estimated annual operating expense would only be about $300,000 a year. The argument, which used a possible $200,000-a-year savings to justify the then $50 million estimated cost of the proposed bridge, was not very persuasive.[5]

The reason the bridge project was so easy to challenge was because it would be built in a location far removed from most of the state's voters. In 1950, approximately 95 percent of the state's population lived in the Lower Peninsula, most residing far south of the proposed bridge site. Approximately 302,000 people lived in the Upper Peninsula. In contrast, the population of Michigan's Lower Peninsula in 1950 was a bit more than six million.[6] Spending money to create a tourist boom in the sparsely populated northern part of the Lower Peninsula and in the equally sparsely populated Upper Peninsula would never be as politically popular as spending money to improve the road system in the densely populated Detroit area.

Although the project had, at best, lukewarm support in many Democratic voting areas, Williams was committed to building a bridge. Republicans eventually acknowledged that the dramatic increase in the number of cars crossing the straits did create a legitimate complaint about the quality of the post–World War II ferry service. But the Republican state highway commissioner resolved the issue, for Republicans at least, by building a new and larger straits ferry, the *Vacationland*.[7] When the ship went into service in January 1952 the superintendent of the state ferries offered the opinion that the past delays drivers had experienced at the straits were over, saying, "I'll eat my hat if there is a line-up of cars at the Straits next season."[8]

Williams, however, did not believe a new ferry was sufficient. To generate public support for building a bridge Williams created the Inter-Peninsula Communications Commission, which was given the job of demonstrating why the bridge was essential. But the commission failed to capture the public's attention and was either ignored by Republican legislators or, worse, made the butt of Republican jokes.[9] Williams realized he needed to recast the bridge project from a partisan campaign promise to a bipartisan proposal. It was a tall order, and the person Williams asked to make it happen was W. Stewart Woodfill.[10] Woodfill was a Republican, but one of the few Republicans who supported building the bridge. Woodfill believed anything that made driving to or around the straits easier was good for straits-area tourism, thus he supported constructing a bridge at the Straits of Mackinac.[11]

Woodfill's support was originally quite casual. The project was something he had a kind word for, but not an idea he was willing to invest much time or energy in supporting. That changed in the autumn of 1949 when Woodfill, in his own words, began working on the bridge project "in earnest." He organized the Mackinac Bridge Citizen's Committee, the bipartisan organization Williams sought. Woodfill privately acknowledged the committee consisted of one Republican, himself, and nine Democrats. Woodfill never convened a committee meeting, instead promising he would do all the work if the others lent their names to the effort.[12] They did, and work Woodfill did.

A bill had been introduced into the legislature's 1950 session to recreate the Mackinac Bridge Authority (MBA), which had originally been established in 1934 and was abolished by the legislature in 1947.[13] To see that it became law, Woodfill was in Lansing weekly, lobbying his fellow Republicans. Woodfill also asked Williams to "lay low" regarding the pending legislation, to avoid partisan attacks on the project. It was advice

Williams followed.[14] In addition to his lobbying in Lansing, Woodfill regularly traveled "the roast chicken and mashed potatoes circuit" to promote the bridge to any group that would invite him to speak.[15]

Woodfill's relentless traveling and lobbying, as well as Williams quiet work to ensure Democratic support, resulted in the bill easily passing the legislature. The House voted in favor of the bill 87 to 1, while the vote in the Senate was unanimous.[16] Shortly after the Mackinac Bridge was dedicated, the *Detroit Free Press* ran an editorial discussing who should get credit for the bridge. The paper agreed with the Greater Detroit Board of Commerce, which had said that while Governor Williams deserved much praise, "behind all that, save a bushel or two of credit for W. Stewart Woodfill, owner of the Grand Hotel on gemlike Mackinac Island."[17] There would be other legislative battles over the bridge, but the battle over the MBA bill forged a critical link between the governor and W. Stewart Woodfill that would eventually change the course of tourism on the island.

Why Williams so deeply cared about the fate of Mackinac Island is impossible to say, but perhaps it is enough to know that Williams and his family regularly summered at the Governor's Mansion and fell in love with the island. Near the end of his time as governor, the family purchased a three-story, fourteen-room cottage to which they came year after year.[18] His attachment to the island was so great, Williams requested his remains be buried on the island.[19]

Because of his interest in the island, Williams became far more involved in the work of the park commission than any of his predecessors. His passion was noted by observers like Harvey Campbell, longtime president of the Greater Detroit Board of Commerce and for decades considered one of Detroit's leading citizens, who wrote to Lawrence Farrell, William's executive secretary, that "no governor that I can remember has ever given a sufficient whoop about the island visitors to raise enough hell to make them happier and more comfortable."[20] Williams was not only willing to raise hell, he eventually envisioned a future for the island which expanded upon the ideas put forward by the Park and Harbor Commission and which led to the Old Mackinac Lives Again weekend of 1941. But this idea took time to grow in Williams's mind. His initial goal was political: to gain control over what he saw as a rogue Mackinac Island State Park Commission presided over by a notorious troublemaker. Williams may not have known exactly what he wanted to do on Mackinac Island, but he did know that to do anything, Commissioner Bill Doyle had to go.

Williams's campaign against Doyle began during the 1948 election. Williams claimed his Republican opponent, incumbent Governor Kim Sigler, "had made his peace with lobbyists and is letting them shape his administrative agenda." To prove the claim, Williams pointed to Sigler's actions regarding lobbyist Doyle. In 1946, Sigler had publicly promised to remove Doyle from the Mackinac Island Park Commission and had done so. But in 1948, Sigler had reappointed Doyle to the commission. Williams claimed it was done to pacify an important lobbyist.[21]

After Williams won the election, he had the opportunity to appoint three new park commissioners, replacing members whose terms were expiring. Williams replaced the three, including then commission chair Louis Simon, with three Democrats. This action created an unintended consequence. Doyle, whose term was ongoing, was the commission's vice chair. In Simon's absence he assumed the leadership of the commission. Williams had a plan to address this. At the next commission meeting, Williams's newly appointed members would call for the election of a new chair. Presuming the vote would divide along partisan lines, with the three remaining Republicans on the board voting for Doyle and the three newly appointed Democrats voting against him, Williams, as an ex officio member of the commission, would cast the deciding vote, removing Doyle as chair.[22]

When the commission met on August 29, 1949, Williams and his three new appointees arrived first. The three Republican commissioners walked into the room together, wearing identical white suits, hats, and each sporting a walking stick. Doyle was obviously up to something. When Chaiman Doyle convened the meeting, things initially went as anticipated. As everyone anticipated, a motion was made and seconded by two newly appointed commissioners to elect a new chair. As everyone anticipated, the vote ended in a three-to-three tie. As everyone anticipated, Williams cast the deciding vote to remove Doyle as chair. What was not anticipated was that after Williams voted, Doyle, still technically the chair of the commission until the vote was recorded, refused to accept Williams's vote. Claiming that no governor had ever voted at a commission meeting, Doyle announced that before the governor's vote could be accepted the chair needed legal advice on whether the governor, as an ex officio member of the commission, was entitled to vote. Until he obtained legal counsel, Doyle refused to officially accept or record William's vote. Having successfully put aside Williams's troublesome vote, Doyle then ruled the motion lacked a majority and thus failed, leaving him as the commission's chair.[23]

A likely furious Williams obtained a ruling from the state's attorney general that he could vote at commission meetings. Although in Michigan, a ruling by the attorney general is considered to have the force of law within state government agencies unless overturned by a court, Doyle wrote his own legal opinion in which he concluded that the governor did not have a vote.[24] To avoid having to face the possibility of Williams using the attorney general's ruling against him, Doyle announced he would not call any meetings of the commission until December, when the law required the commission to meet, unless Williams agreed that the matter of the governor's vote would not be discussed.[25] Williams did not agree, and so the commission did not meet.

A donnybrook was expected when the commission met in December, 1949. Instead, when the commission gathered together, Williams was absent. As had happened at the commission's last meeting in August, the commission deadlocked along partisan lines regarding the election of a new chair.[26] Doyle again ruled that the motion had failed and thus he remained in the office. Not only did Williams decline to do battle, in December 1950, Williams allowed Democrat James P. Dunnigan to cast the deciding vote allowing Doyle to finally become the elected chair of the commission.[27]

Why Williams had a change of heart regarding Doyle was never explained. In December 1949, Williams explained his failure to attend the commission meeting by saying, "I do not choose to suspend the important business of the State at this time to deal with this relatively unimportant issue." The *Detroit Free Press* didn't believe it. The paper concluded, "G. Mennen Williams joined a long and honorable line—governors who have bowed to W.F. Doyle."[28] Why in 1950 Williams allowed Dunnigan to cast the deciding vote electing Doyle chair also went unexplained.

An intriguing, but undocumented, hypothesis is that like Kim Sigler before him, Williams cut a deal with Doyle, perhaps brokered by Woodfill.

In the fall of 1949, W. Stewart Woodfill was hard at work seeking to pass legislation to recreate the Mackinac Bridge Authority (MBA), something Williams cared about deeply. Passing the bill was likely going to be a hard battle in the Republican-controlled legislature that only three years earlier had voted to abolish the original MBA.[29] To win the battle, Woodfill may have proposed a grand bargain. Doyle, one of the capitol's most effective lobbyists, and a good friend of Mackinac Island, would quietly lobby for the MBA. In return Williams, would abandon his effort to remove Doyle as chair of the Mackinac Island State Park Commission. It is the kind of bargain the pragmatic Woodfill might propose, Doyle might accept, and Williams might embrace to obtain a major legislative goal.[30] The logic behind such a deal makes for compelling speculation, but there is no evidence to support the claim. It's hard to believe, however, that Williams bowed to Doyle without getting something important in return, and the reauthorization of the Mackinac Bridge Authority would be a very important item on the governor's agenda.[31]

But if they had agreed to such a bargain, Williams and Doyle did not appear happy about it. A gossip column in a 1951 Detroit newspaper included the interesting news that both Williams and Doyle had been observed reading newspapers in the Lansing Y Athletic Club, sitting on opposite sides of the room in which the papers were kept. "The silence in the room was deafening and the temperature sub zero."[32] The relationship between the two men was not improved by rumors that Doyle might run for either governor or lieutenant governor on the 1952 Republican ticket.[33]

In 1953, the conflict regarding the Mackinac Island State Park Commission between Williams and Doyle was finally resolved. Doyle, his term on the commission ending, announced he was not a candidate for reappointment and would resign immediately as commission chair, further agreeing that his replacement would be Williams's close ally James P. Dunnigan.[34] But before leaving the commission, he engineered passage of a bill reformulating the park commission's membership using what newspapers called the "Bill Doyle formula."[35] The commission would be composed of six members, with no more than three from one political party. The resident commissioner was retained on the commission, but the governor could appoint any island resident to the seat rather than the individual nominated by the city's mayor. Finally, the law abolished the governor's ex officio standing on the commission by omitting any reference to it.[36]

The new law, coupled with Doyle's announcement that he would resign immediately as chair and did not wish reappointment to the commission, allowed both men to save face. Doyle avoided the public humiliation of Williams almost certainly passing him by and appointing someone else commissioner. But he also got something. He could claim he had defeated Williams over the question of whether or not the governor could vote at commission meetings. The Doyle formula had made the question mute by eliminating the governor as an ex officio commission member.

From Williams's viewpoint, he had finally removed Doyle as commission chair. He had also avoided what might have been a nasty partisan fight if he appointed anyone but Doyle to the commission. The law authorized the governor to nominate anyone to the commission, but the governor's appointee was subject to Senate confirmation. Normally, confirmation of a commission appointment was routine; however, if Doyle asked his many friends in the Republican-controlled Senate to make trouble, the governor would be in a very difficult situation.[37] The matter was resolved, but Doyle never forgave Williams for ousting him from the Mackinac Island State Park Commission. At the time, Doyle said, "He [Williams] replaced me for one reason and only one, . . . I have beaten him at every turn of the wheel and he just can't take it."[38]

With Dunnigan running the commission, in 1955, Williams began what he hoped would be the reorientation of the entire region's tourist industry toward history. He attempted to accomplish this by forming a new committee, composed of himself, the members of the Mackinac Island Park Commission, the mayor of the island, the members of the Michigan Historical Commission, and several university professors. Williams attended the group's first meeting and shared his hopes for an aggressive regional historical agenda that would encompass Fort Mackinac, Mackinaw City, St. Ignace, and Sault Ste. Marie. Privately, Williams gave Dunnigan the job of implementing this vision.

> Since you are Chairman of the Commission, Jim [Dunnigan], I would appreciate you setting the wheels in motion and contacting the people outside of

> the Commission who I have listed here and perhaps others and develop with them some long range and well as a short range program to permit the people of Michigan and the rest of the country to enjoy our rich heritage to the full.[39]

Dunnigan, who had a reputation for getting things done, seemed a good choice to move Williams's plans forward.[40]

Dunnigan, however, did not live up to Williams's expectations. Dunnigan was unable to construct the coalition Williams had envisioned. Worse, he quickly reignited old conflicts and made new enemies. Dunnigan and the commission entered into a bitter fight with the island's city government over the fate of one of the island's most historic buildings.[41] Dunnigan was equally unable to resolve conflict with one of the commission's key constituents. The Michigan State Medical Society had restored the historic Beaumont Memorial and, when the project was completed, given it to the park commission. But following the gift, the relationship between the society and the commission had gone from bad to worse, eventually becoming so poisoned that by the middle of 1957, the society was asking the governor to remove the property from the Mackinac Island State Park Commission's portfolio and make it the responsibility of the Michigan Historical Commission.[42]

Dunnigan also failed to establish a successful working relationship with other state agencies. For example, the Michigan Historical Commission had begun a program to install historical markers around the state. It identified several markers it wished to be placed on Mackinac Island and sought to partner with the Mackinac Island State Park Commission in completing the work. But the partnership faltered when the Mackinac Island Park Commission threw up a roadblock because the historical commission's proposed wording for the markers did not acknowledge the markers were on land administered by the park commission.[43]

The number of words that could be placed on a marker were few. Saying who owned the property would necessarily mean fewer words about why the site mattered. Furthermore, giving the commission this recognition opened the door for every organization that held title to a historic site where a marker might be placed to demand the same recognition. It was a parochial demand and governor's staff placed the blame for the situation one man—Dunnigan. As one staff member put it in a letter, "Jim Dunnigan has his Irish up."[44] Eventually a compromise was worked out. The wording remained as the historical commission desired, but the seal of the park commission was added to the bottom of the marker.[45] But Dunnigan's "Irish" was becoming a problem.

Eventually Williams came to the conclusion that despite Dunnigan's loyalty, he could not push forward the governor's agenda. To accomplish that, Williams turned to an old ally, W. Stewart Woodfill. Woodfill told a dramatic tale about how Williams came banging on his door in the middle of the night, sat on the corner of Woodfill's bed, told him about his plans for the island, and asked Woodfill to join the commission.[46] Though the story sounds a bit too good to be true, Williams never denied it, and it is certain that he did ask Woodfill to become a park commissioner. Persuading Woodfill to join the commission took some effort, as Woodfill was known to routinely turn down requests for his time. For example, in 1955, he had declined election as president of the Chamber of Commerce.[47]

Woodfill, however, had his own reasons to think seriously about the governor's proposals. Woodfill knew that the tourist business on the island was growing, but he also knew the number of tourists passing through the straits region was growing far faster than the number of people visiting Mackinac Island. This trend threatened to eventually harm the tourist trade on Mackinac Island. Woodfill had, in the 1930s, thought about using history to increase the number of visitors to the island. Williams may have reminded him of those thoughts, and of the possibility of profiting from embracing the island's historical status, making it an antique, but a cherished and much loved antique. Woodfill may have seen a vision that not only benefited the island, but also fit well with his own ideas about the place of the Grand Hotel within the tourist industry. He was, however, cautious enough to condition his agreeing to join the commission on two points. Woodfill must be made the commission's chair, and Williams would give his unequivocal support to Woodfill's decisions.[48] Williams agreed, and Woodfill attended his first meeting as a commission member on June 22, 1957, but with Dunnigan chairing the meeting.[49]

James P. Dunnigan and W. Stewart Woodfill both had considerable egos. Dunnigan knew that Williams had agreed to make Woodfill chair. Neil Staebler, chair of the Michigan Democratic Party, seems to have broken the news to Dunnigan, writing to Williams,

> Jim is agreeable to have Woodfill designated as chairman of the Commission and will support a change in the Commission's rules to accomplish this. He understands that Woodfill will lead a determined effort to rehabilitate the Island and will join wholeheartedly in the program.[50]

To avoid this change embarrassing Dunnigan "in politics and business" Woodfill agreed to postpone becoming chair, agreeing that he would assume the office "when the matter would be given no press and there be no fanfare about it."[51] Woodfill saw this happening toward the end of the summer. Dunnigan did not. He reminded Woodfill that "the law" was specific in that it called for the park commission to elect its officers at the commission's annual meeting held in December.[52] Although the two men were talking a difference of only a few months, this became the source of a bitter disagreement.

Woodfill placed Williams squarely in the middle of the conflict when he wrote to Williams regarding the "embarrassing confusion" that existed and suggested, "It is my opinion that upon your request, Jim would by pleased to resign on July 8th [the next scheduled meeting of the commission] and nominate me for the Chairmanship at that time." Woodfill knew full well, as did the governor, that "Jim" would not be "pleased to resign on July 8th," but Woodfill pressed for the immediate resolution of the matter by making a threat. "The only way that I can see to avoid these embarrassments now is that the matter be cleared up happily on July 8th by my election as Chairman or that I quietly resign my membership in the Commission."[53]

In the end, the governor managed to resolve the matter in a way that allowed both men to walk away with their pride intact. On August 12, 1957. Dunnigan resigned as park commission chair, although not from the commission itself, and asked the body to elect Woodfill to replace him.[54] Williams was wise enough to soften the blow by making sure Woodfill was on his best behavior. In the words of the commission's minutes, Woodfill spoke "most feelingly" about the fine work Dunnigan had accomplished. Woodfill did not like nor trust Dunnigan and unless prodded to say something nice about him would most likely have spoken "most feelingly" about his own goals, not his predecessor's accomplishments.[55] Woodfill, in fact, did not like any of the commissioners, regarding each of them with an opinion that ranged from indifference to contempt.[56]

Mackinac Island State Park Commission Meeting, ca. 1958. *Far right and counterclockwise around the table*: W. Stewart Woodfill, Joseph H. Thompson, Eugene T. Petersen, Mariana B. Rudolph, Walter J. Murray, James P. Dunnigan, Carl A. Nordberg, Neil Downing, and W. F. Doyle.
COURTESY MACKINAC ISLAND STATE PARK COMMISSION.

In addition to ensuring that Woodfill praised Dunnigan's work, Williams found another way to assuage Dunnigan's wounded pride. Dunnigan would reminisce that Williams gave him a choice of chairing one of two state bodies. At the time, it was illegal to chair more than one state commission or board, thus Dunnigan could pick only one. Williams offered Dunnigan the opportunity to chair the Mackinac Island State Park Commission or chair the State Highway Commission Advisory Board. Dunnigan's business was selling road asphalt. As he would recall, "For obvious business reasons I chose the Highway Chairmanship." Dunnigan, however, was honest enough to also acknowledge the politics of the situation, adding in the same recollection, "An embarrassment was avoided for all parties since I felt that Williams wanted Woodfill to head up the ongoing restoration program and I had freed his hands."[57]

At the commission meeting of August 12, after its new chair was in place and Dunnigan had excused himself to attend to other business, the governor entered the room and joined the meeting. What he said was likely very similar to what he had told Woodfill when he asked him to join the commission. As the minutes of the meeting state, "Governor Williams then expressed the feeling that more attention should be given to the historical background of the Island and its history."[58] Although Williams talked about other projects, such as improving the trails and footpaths to better enjoy the island, it was clear to everyone that history was William's primary concern. As one Detroit newspaper would reflect about a year later, as the first fruits of Williams dream were becoming apparent, "Commission members visualize building Mackinac Island and the straits area into a tourist attraction rivaling Colonial Williamsburg, Va."[59]

Woodfill would later reflect:

> Now as to who should have the credit for the development of the Mackinac Island Park Commission from a palace guard group that was long dormant into a very successful operation for the state of Michigan, it is G. Mennen Williams. It was because he had the unusual political courage to select a person whom he considered a good businessman and one whom he respected as to the integrity and motivation even though I was of the opposite political party and had never subscribed one dime to any of his campaigns and never voted for him and never proposed to do so and he knew it. I told him so. His was one of the greatest, if not the greatest act of statesmanship that I have ever known of any political person.[60]

Williams had pointed the park in a fundamentally new direction. Woodfill signed on to drive forward the governor's agenda. It would be a challenging project. Eugene Petersen, who would be fundamental to the successful implementation of the governor's ideas, described his first visit to Mackinac Island in 1957 this way:

> It always bothers me that at many historic sites you have to run a battery of gift shops before the main attraction; however, on Mackinac Island this is only the continuation of a tradition. Instead of beads for beaver belts, there are fudge, post cards, and film for cash.[61]

Petersen may have been disturbed by the commercialism that obscured the island's history, but in criticizing it, he placed his finger on something needed to make the governor's dreams come true: cash. Redirecting the island's tourist industry toward history would require money, and a lot of it. The money could come from the state legislature, and Woodfill asked the legislature to supply it. The state had been giving the commission about $100,000 a year. Woodfill thought $500,000 a year would allow "a limited start of the long-term program of improvements," with $250,000 a year being sufficient after the improvements were completed.[62] A half million dollar request was introduced in the legislature. As usually happened with commission requests for generous funding, the money was not appropriated.[63]

Woodfill and Williams likely expected this and had prepared an alternative plan, raising money by selling bonds. The commission, the governor, and others had been talking about finding a way to utilize bonds since 1955 and had looked closely at the island's Parks and Harbor Initiative of 1941.[64] Woodfill turned this discussion into a legislative proposal by having his personal attorney draft a law that, if passed, would authorize the commission to sell bonds, paying for them by charging admission fees to many of the park's attractions, including Fort Mackinac. Woodfill, who ran the Grand Hotel as if it was a personal kingdom, included in the bill similar independence for the commission from the rules applying to state agencies. Besides authorizing the commission to sell bonds, the proposed legislation gave the commission power to build, restore, or reconstruct buildings as it wished and ignore the state's civil service laws when hiring employees. It also allowed for an independent accounting system that would not be required to follow standard state accounting procedures. If passed, the bill would allow the Mackinac Island State Park Commission to do things its way, or, more properly, Chairman Woodfill's way.

Woodfill had his bill, but he realized it was unlikely that the Democratic governor could successfully shepherd it through the Republican-controlled state legislature. Woodfill, however, knew who could accomplish the feat: Bill Doyle. By 1958, the two men had come to an understanding over their past differences that allowed them to work together. It may have happened in 1949 when the MBA bill became law, but how, and when, it had happened no one can say with certainty. R. D. "Dan" Musser, Woodfill's nephew, would explain the reconciliation this way:

> They finally reached a truce walking down a street downtown. Mr. Woodfill used to walk the street three times a week just to see what was going on. They ran into each other and I don't know who said it first, but I think they both would say they did, but they made a deal that let's call a truce.[65]

However it happened, the two men were again talking to one another, and Woodfill turned to Doyle to steer the bill through the 1958 session of the legislature.[66] In addition to acquiring Doyle's help, Woodfill likely reprised his role as master publicist for the bridge campaign in support for the bond legislation. One fruit of Woodfill's labor was likely found in the *Detroit Free Press*, which wrote editorially that the bill "would be a boon to all of Michigan." What the *Detroit Free Press* seemed to particularly like was that this boon "would cost the taxpayers nothing."[67] The bill passed early in 1958.[68] The new law gave Woodfill and the park commission something very close to a blank check to do whatever it wanted, so long as the commission made the check good through admission fees.[69]

With the law signed, Woodfill and the governor wanted work to begin quickly, but two problems stood in the way of this happening. First, because of various legal procedures required to sell bonds, the money from a bond sale would not become available until the fall of 1958. To solve the problem, Woodfill advanced the commission the money needed to begin work. Although everyone expected there would eventually be enough revenue to pay him back, everyone realized that their expectations could be wrong. In a carefully worded agreement between Woodfill and the commission, it was stated that it was the commission's intention to repay Woodfill's loan from admission charges to be created, but it was also agreed that should the situation arise that the commission's revenue from new admission charges not be sufficient to pay the loan, the money would be a gift from Woodfill to the park.[70]

With the money in hand, Woodfill faced a second problem, he lacked the staff to spend it. He needed someone who could construct a historically themed park and no one then on the park staff fit the bill. To get such a person, Woodfill borrowed Eugene Petersen, the director of the state historical museum in Lansing, from the Michigan Historical Commission. It was a tremendous gamble. On paper, Petersen had good credentials. He had received a doctorate in history from the University of Michigan in 1952. He had taught history briefly but soon became director of the State Historical Museum in Lansing, a post he had held for several years.[71] But the state history museum, like the Mackinac Island State Parks, was chronically underfunded.[72] While leading it, Petersen had never done anything like what Woodfill was asking of him. Nor was Petersen's personality much like Woodfill's or the other park commissioners.' In one of his first public appearances at a commission meeting, a newspaper reported, "the quiet, scholarly young man seemed a little out of place among the cock-sure extroverts seated at the long table."[73]

Chairman Woodfill made Petersen's first priority rehabilitating Fort Mackinac. Woodfill wanted results immediately, before the park opened in 1958, and Petersen delivered. He removed an exhibit of carriages from the Soldier's Barracks and installed a chronological exhibit about the fort. Lacking historic artifacts to display, Petersen relied on murals and dioramas that employed large and striking graphics.[74] It was impressive and historically accurate. It was also a quick fix. Petersen knew he wanted and needed more.

Woodfill had a somewhat different point of view. Woodfill often talked about his love of Mackinac history, and presented himself to newspaper reporters as "an ardent student of the Island's lore and historical heritage."[75] But Woodfill was also a businessman, and Petersen realized that for Woodfill, business came first. Said Woodfill, "Remember, our money largely comes from the uneducated. We want an authentic restoration the critical historian can enjoy, but we also must have mousetrap gimmicks to pay the load."[76]

Woodfill was changing the park commission's business model into one that required the public to bankroll Peterson's historical displays. To make this happen, Woodfill applied the successful business model of the Grand Hotel to the park. At the Grand, business relied on high standards, exemplary service, and a willingness to exaggerate or outright lie in public relations statements to entice customers to visit. In the park, the high standards were represented by Petersen's well-researched and equally well-constructed exhibits. The exemplary customer service was delivered by generally well-trained and usually personable historical reenactors and other public-facing staff. As for gimmicks to appeal to the uneducated, Woodfill latched onto the idea of an exhibit regarding military punishment. A flogging pole had stood in the fort. Woodfill not only wanted the pole quickly reconstructed, he wanted it augmented with a suffering mannequin, likely showing the scars of a recent whipping. "I predict this sort of stuff will be the best publicity and attraction we will have."[77] The idea had little appeal to Petersen, but at least this was a truthful gimmick unlike some of the stories Woodfill spun about the Grand Hotel. And when he thought about it, Petersen recognized that while he believed that good history was critical, making enough money to keep the doors open was also essential.[78]

Woodfill improved the odds of making money by building a new entrance to Fort Mackinac. For as long as visitors came to the island, they had trudged up the steep embankment from the harbor to the fort's front gate. It was a strenuous climb that not everyone was willing to make. In the first years of the 1950s, the commission had discussed the possibility of building an escalator, or perhaps a cable car, to take people up the hill.[79] Woodfill's solution was far less expensive and far less jarring to the island's historical ambiance. In 1958, the "Avenue of the Flags," at the rear of the fort, was opened. It was located at a new, mandatory, carriage tour stop.[80]

Visitors could alight from their carriage, pay their admission fee, and walk downhill into the fort by descending a staircase connecting the fort's rear stockade to the parade ground.[81] Woodfill also helped ensure success by pressuring the carriage tour operators to not just stop at the Avenue of the Flags but also to talk up the fort's historical reconstruction project. As Petersen wrote to Woodfill, "Back in 1958, as you well know, we urged and coerced them [the carriage tour operators] into promoting Fort Mackinac by demanding that the drivers sell the historical restoration program, and in due time they did."[82]

Whether it was Petersen's masterful implementation of public history or Woodfill's employment of gimmicks, or, more likely, some combination of the two, money flowed into the commission's coffers. When admission had been free, about 100,000 people had visited Fort Mackinac each year. There was concern what would happen to attendance when the

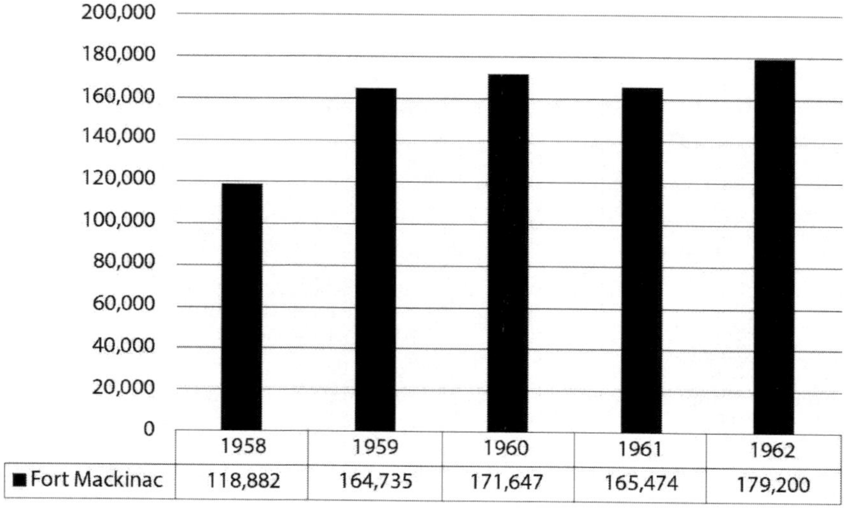

Fort Mackinac attendance, 1958 to 1962.
SOURCE: FIGURES COURTESY MACKINAC ISLAND STATE PARK COMMISSION.

admission fee needed to pay back the bonds, originally set at fifty cents for adults, was charged. To break even in 1958, the commission needed approximately 58,000 customers to visit the fort. To everyone's delight, when the fort closed for the season, approximately 118,000 customers had paid to visit.[83] Over the next several years, attendance plateaued but remained well above what was needed to pay back the bonds.[84]

Petersen continued to develop serious historical exhibits on the island and within a few years, Woodfill and Petersen had made real the hopes G. Mennen Williams had expressed before the commission in August 1957. But when Williams chose not to run for reelection in 1960, a critical moment was reached. A year after Williams left office, Woodfill resigned as commission chair.[85] His accomplishments did not go unrecognized. At the meeting at which Walter J. Murray was elected the new chair, Woodfill received a special award of merit from the American Association of State and Local History "for his outstanding contribution to the cause of history and particularly his leadership in the program to restore Fort Mackinac, Fort Michilimackinac, and other historic sites in northern Michigan."[86]

Woodfill had accomplished a great deal but what he had not done was put into place guarantees that would secure those accomplishments.

Securing and continuing what had been achieved was the task left to Petersen, who was dismayed when Woodfill stepped down from the chair. His dismay was based on hard truths about Mackinac Island that he had learned, and which did not bode well for the future.

> Beneath the charm of Mackinac Island's historical and natural beauty was a political underground that resisted change or, worse, was indifferent to it. Vested interests with a large financial stake in serving the thousands of summer visitors welcomed our effort at Fort Mackinac, but only if it didn't affect their narrow interests. Commissioners who supported the bond-financed museum program did little to secure funding for related park improvements and seemed to be satisfied with Carl's [Nordberg, park superintendent] woefully inadequate budget. Some of them obviously relished controversy with the Island businessmen and basked in the social recognition and power of running the park.[87]

It was a fraught moment when everything that had been done by Williams, Woodfill, and Petersen hung in the balance.

CHAPTER 7

The Director, the Mayor, and History, 1961–1990

W. Stewart Woodfill and Eugene Petersen had begun transforming the Mackinac Island State Park into an institution centered on historic tourism. But they had done it largely by themselves, empowered by the unwavering support of Governor G. Mennen Williams. In 1961, Michigan had a new governor. As a result, Woodfill, a man who in Petersen's words, was "ill-suited to work in state government," and who "in another age... would have supported a benevolent dictatorship,"[1] soon resigned as commission chair. Petersen was left to carry on promoting the cause of historic tourism within the park. This job would have been challenging under any circumstance, but it was made even more challenging by the reappointment of Bill Doyle as a commissioner.

When Governor Williams had finally pushed Doyle off the commission, he likely expected Doyle would fade away. Doyle, however, had seen governors come and governors go. He had been kicked off the commission in the past and nevertheless found his way back onto the commission. When Williams did not run for reelection in 1960, Doyle was soon speaking about being reappointed to the commission with William's successor, Governor John Swainson.

How Doyle persuaded Swainson to appoint him is unknown. In a dismissive mood, Woodfill wrote:

> My guess is that Doyle has been working with great diligence licking the Governor's boots, and working to get this appointment, and probably has it made. He likely used our efforts to get legislative funds for the Governor's Residence as one of the shoe-horns to get the appointment, for he has kept the Governors' office regularly informed of *his* great work to get the funds to fix up the house.[2]

Although Woodfill was keen to make work on the Governor's Residence the result of "our" efforts, meaning his own, Doyle likely had more to do with getting the legislature to appropriate the money to "fix up" the governor's island residence than had Woodfill. In a less dismissive mood, Woodfill knew Doyle could offer the governor something more important than a repainted bedroom. As he likely had done with other governors, Doyle may have promised his help in getting the governor's agenda through the legislature.[3] Whatever the reason, Governor Swainson reappointed Doyle to the Mackinac Island State Park Commission in 1961.[4]

Doyle's reappointment introduced a wildcard into the park. Early on, Commissioner Doyle went out of his way to offer an olive branch to Woodfill. Speaking of what had been done during Woodfill's chairmanship, Doyle said, "The big thing before the Commission now is the improvement and the extension of the splendid restoration program which got underway in 1958." For those who wanted to believe that old dogs could learn new tricks, Doyle's generosity in describing Woodfill's efforts as a "splendid restoration program" may have offered hope. Woodfill, however, likely accepted Doyle's compliment as his just due and waited to see exactly what Doyle had in mind regarding "the improvement and the extension" of Woodfill had begun.[5] Doyle was a loose cannon, and Petersen would reminisce years later that Doyle was the commissioner who gave him the most trouble.[6]

Petersen perhaps thought that Doyle's important suggestions would focus the future of the park's newly created historical program. He was wrong. Petersen learned that the commissioners, including Doyle, weren't much interested in the park's historical program. As long it remained popular with the public and brought in sufficient revenue to pay for the bonds that funded it, the commissioners generally let Petersen proceed as he thought best.[7] What forced the commissioners to seriously think about the future was a job offer to Petersen.

Petersen accepted a job in 1966 at the Henry Ford Museum in Dearborn, Michigan. When Woodfill hired Petersen, he had Petersen report directly to him, as the commission chair. Everyone else who worked for the park reported to park superintendent Carl Nordberg, who had held that position since 1945. Woodfill himself would later realize he had created a two-headed monster. As long as Woodfill was chair, with both Petersen and Nordberg reporting to him, things worked reasonably well. Woodfill would reconcile any conflicts, generally in Petersen's favor. But after Woodfill resigned as chair things began to unravel. Petersen grew frustrated with what he perceived as a lack of support shown for the history program by Nordberg. Petersen feared that to resolve the problem of an independent history program the commissioners would eventually make Nordberg, who was the boss of everyone else, his boss.[8]

It is likely Nordberg very much would have liked that. Petersen reminisced that in one meeting between the two men, Petersen told Nordberg that the historical program was as important as the rest of the park and if Nordberg didn't support it as Petersen felt was necessary, Petersen would take his case directly to the commissioners. Who could blame Nordberg for wanting to rationalize the park's administrative structure under his control, and incidentally put Petersen in his place? The commissioners, however, ignored the tension between the two men. The two-headed monster continued for years, only being addressed in 1966 when Petersen informed them that he was leaving.[9]

Petersen submitted his resignation, but the commissioners refused to accept it. In this highly irregular situation, Petersen made two demands that would persuade him to stay. First, the commission must rationalize park administration by creating a new senior management position, the director, who would have oversight over all ongoing park operations. Second, the commission must appoint him the first director. Either Petersen would be in charge, or he would leave. The decision of who would be in charge, Petersen or Nordberg, would inevitably lead the park in very different ways.

Petersen held a doctorate in history from the University of Michigan.[10] W. Stewart Woodfill gave him the opportunity to develop a major historic site. It was a job Petersen enjoyed and at which he excelled. Nordberg was a 1928 graduate of Michigan State University, where he had been the quarterback on the school's football team. His college athletic experience led to a successful career coaching high school football.[11] In 1937, Nordberg began working during the summer on Mackinac Island, originally as the dockmaster. Bill Doyle liked what he saw and in 1945 arranged for Nordberg to become park superintendent.[12] His long tenure as superintendent made it clear he could manage the park as it had been run since the turn of the century, successfully manage the park staff, and successfully cope with the often challenging personalities of the various commissioners. But, as Petersen liked to point out, he was not particularly interested in the history.[13]

Petersen later recalled that "without much discussion Carl Nordberg became park manager under my direction, and I was named Director."[14] Petersen's reminiscence accurately reflects what happened at the commission meeting that accepted his demands and appointed him director, but it does not reflect the commission politics that led to the decision. There were members on the board, led by Doyle, who had a high opinion of Nordberg, or at least found him the man with whom they liked to work.[15] Doyle repeatedly complained that Petersen was a "bureaucrat" and more importantly was "one who will not take orders,"[16] most likely orders issued by Doyle himself. Doyle, and everyone else on the commission, understood that Nordberg was "his" man, while Petersen was the protégé of Woodfill. The board unanimously appointed Petersen director. Unanimity, however, was achieved by downplaying the importance of the decision.

Woodfill and Doyle both claimed that the decision was a routine one. Woodfill said, "The only difference I can see under this reorganization plan is that Mr. Nordberg will report to Dr. Petersen rather than directly to the Commission." Woodfill knew this decision was about much more, but he chose not to talk about it. If Woodfill was uncharacteristically taciturn, Doyle's public statements bordered on the ridiculous. Doyle announced, "This action is in line with Governor [George] Romney's policy of streamlining agencies of the executive branch of government as we understand it." The idea of Doyle complying with a governor's wish that did not serve his own agenda was laughable given his long history of disrespecting the wishes of several previous Michigan governors. So too was Doyle's statement that "Good common sense tells us . . . there must be one administrative head only."[17] If it was just a matter of good common sense, why did it take the commission so many years to realize it? Events had forced a decision upon the commission, Doyle lacked the votes to retain his man in charge, and as a result, historical tourism became the park's major focus.

The park commissioner changed the way tourists experienced the park, but they were far from willing to change the way they thought about themselves. In many ways, the commissioners during the 1960s and 1970s continued to view themselves much as had the people who held the post prior to 1931. This sensibility created a serious public relations problem that could have detrimentally affected public support for Petersen's work. Conflict of interest standards was one example of where the commissioners badly lagged behind the times, dragging themselves, and indirectly the park, into a very difficult situation. The issue first exploded over a decision to renew the lease of state land to the Grand Hotel. The land was used as part of the hotel's golf course.

The lease renewal had been controversial within the commission itself. Commissioner James P. Dunnigan labeled Woodfill a "promoter" over the matter, while Woodfill said Dunnigan's opposition was personal. Commissioner Walter J. Murray shared with the other commissioners that he also believed Dunnigan was carrying on a vendetta against Woodfill. Dunnigan vigorously denied Murray's charges. While the commissioners were engrossed in personal politics, they collectively dismissed a statement by the Michigan attorney general, who wrote, "Should the Commission enter into the proposed lease, . . . there would be a substantial conflict of interest." Rather than heed the attorney general's advice, the commission voted to renew the Grand Hotel's lease of state land.[18] They would pay dearly for the decision.

Under the headline "Island Guardians Waive the Rules," Roger Lane wrote in the *Detroit Free Press*, "As befits its princely estate, the seven-member commission at times has shown royalty's impatience with trifles and annoying regulations suited more for commoners." Lane went on to say that by renewing the golf course lease, the commissioners had ignored "some pretty starchy standards" laid down by the governor that stated, "State employees must avoid all situations were prejudice, bias or opportunity for personal gain could influence their decisions." Lane quoted the governor's directive, noting that state employees should also avoid "that which suggests favoritism or personal gain as the motivating force in the conduct of state government."[19]

But ignoring the governor's broadly worded directive was only part of it. Lane went on to write, "The Commission has now, astonishingly, thumbed its nose at Michigan's attorney general." Lane added that several commissioners rented or leased state-owned land, suggesting that the golf course lease was part of a broader pattern of abuse.[20] Having already thumbed their nose at both the governor and the attorney general, it surprised no one that the island's royalty chose to ignore Lane, as their predecessors in the 1920s had ignored those who opposed the commission's efforts to seize control of much or all of the island's private property.

Approximately a year after Lane's article was published Attorney General Frank Kelley ordered that the six commissioners who held leased state land "explain their interests." To ensure the commissioners addressed his order with urgency, Kelley also announced he was holding up the renewal of fifty-five leases, many for land underneath cottages, issued by the commission to see if the rental price was fair. Kelley was holding the commission's entire leasing process hostage until he received answers. Kelley had had chosen his target carefully. The commissioners knew that the fees for cottage leases were far below market value.

From 1895 to 1933, the commission had used leases to produce revenue and had set rates as high as the market would bear. However, the Depression "shattered" the island's economy.[21] One result was that cottage rents were reduced substantially in the hope that it would keep cottage owners from walking away from their cottage. In the 1960s, the Depression was long gone but the "nominal rent program" remained in place, benefiting the cottage lease holders, including the five commissioners who each leased park land for a cottage they owned.[22]

Woodfill was in an even more challenging situation. The attorney general had issued a conflict of interest standard that stated, "Any state officer . . . who serves as an officer or director of a private corporation doing business with that institution is involved in a substantial conflict of interest contrary to law."[23] The lease of state land to the Grand Hotel clearly violated this standard. As he had in the past, Woodfill ignored his culpability and instead issued a fog of obfuscation to distract the public. He claimed Kelley was engaged in a "vengeful play."[24] Commissioner Doyle supported Woodfill, later calling Kelley's actions a "witchhunt."[25] To further change the topic, Woodfill offered Governor Romney his resignation from the commission.[26] In the short term, the fog of words worked. Instead of worrying about the obvious conflict of interest, Governor Romney focused on talking Woodfill out of resigning. But a year later, in 1969, the governor accepted a second letter of resignation written by Woodfill.[27]

Woodfill was not the only commissioner who found it expedient to resign rather than face legal action by the attorney general. In 1973, Mariana B. Rudolph, another commissioner with business ties that intersected with commission activities, decided after "long thought" to resign because of the appearance of a conflict of interest. Like Woodfill, Rudolph denied there was an actual conflict, but like Woodfill, she preferred not to have the matter taken up in a court.[28] Kelley's review moved very slowly.[29] He may have preferred the club of an ongoing investigation to settling the issue. Over time, the commissioners, who ran the island park as a "princely estate," began to accept a more contemporary understanding of accountability. But it was a slow process.

As the commission wrestled with legal problems largely of their own making, Petersen and the staff were able to demonstrate, in dollars and cents, that historic tourism had been a good choice through which to develop the commission's island property. Between 1958 and 1990, each year, approximately 203,000 persons visited Fort Mackinac, with small but regular increases in attendance over the course of approximately thirty-three years.[30]

The park staff had adopted historic tourism as their basic rationale for park operation. The question remained over how the island's government would react to the commission's choice. As Petersen knew, there were deeply entrenched, narrowly focused interests within the island's tourist industry.[31] Would their elected representatives embrace historic tourism, ignore it, or perhaps even work against it? The history

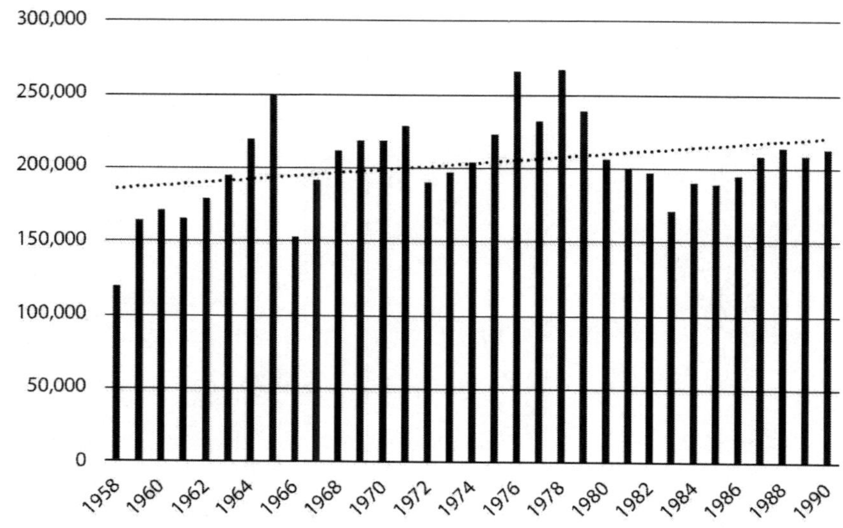

Fort Mackinac attendance, 1958 to 1990.
SOURCE: FIGURES COURTESY MACKINAC ISLAND STATE PARK COMMISSION.

of the relationship between the commission and the island municipal government made any of these choices a possibility. What tipped the balance in favor of historic tourism was likely the continuing growth in the number of people coming to Mackinac. The best argument in favor of historic tourism was that it brought an ever-increasing number of visitors. In 1961, 696,993 tourists visited the island. In 1989, 1,712,368 came.[32] The increase of approximately one million visitors between 1961 and 1989 created a substantial opportunity for increased profit. Simply put, historic tourism made everyone money.

Likely because of the handsome profits being made, city government found the political will to pass not always popular laws protecting and promoting the island's Victorian era architecture. As early as the 1950s, the Michigan Historical Commission had called for maintaining the "historic atmosphere" of the island by maintaining a more Victorian image through zoning restrictions.[33] It was a discussion that took on renewed public interest in the late 1960s when Straits Transit, one of the ferry companies which brought visitors to the island, applied for a building permit for a new dockside terminal. The company proposed building a thirty-by-forty-foot steel-sided structure with a metal roof. Island traditionalists were profoundly upset that visitors' first sight of Mackinac Island's harbor would feature a metal pole barn. W. Stewart Woodfill wrote, "As for the metal building I think it would be a tragedy to permit that to be done." Straits Transit, Woodfill asserted, was simply trying to save some money at the expense of the entire community.[34]

The city council voted against the permit. The city wanted a wood building, one that the president of Straits Transit alleged would cost ten times that of the metal structure. The town's newspaper wrote that while the council's intentions were good, its legal footing was unsound. The paper concluded that there was no city ordinance that required something other than a metal building.[35] But should there be? A year later a member of the Michigan Society of Architecture's board of directors publicly criticized the village government for allowing "honky-tonk architecture" in the downtown area, and worried that unless it was brought under control, the island would see fewer visitors.[36]

For several years, the debate simmered, before again erupting into public view in 1974 when the city council received a request to construct a mini-golf course. The request for that building permit set off a fierce debate. What those opposed to the idea may have feared was something like the mini-golf course that had opened in St. Ignace in 1962. The course named each hole after a famous local tourist attraction, made the hole in some way reminiscent of that attraction, and included brochures at the hole about the attraction. The Grand Hotel was featured at the sixteenth hole, which was also at the top of the five-tiered course. At the Grand Hotel's hole, a player teed up on the hotel pool's diving board, aiming the ball into the drain at the pool's bottom.[37] The fear of such a gawdy attraction led opponents of the mini-golf course to worry that the town might eventually resemble New York City's Coney Island, and to argue that this was not simply a matter of aesthetics, it was bad for business.[38]

Thinking along these lines, council member Tom Pfeiffelmann pointed out that anything could be built, no matter how gaudy. If he wanted to put up a Ferris wheel in the middle of the town, the council would have to give him a license. He mused that a Ferris wheel would have a nice view of the Mackinac Bridge and nearby Round Island and would likely turn a nice profit.[39] Proponents of the permit said the city had no right to regulate the "look and feel" of the community, regardless of its impact on other businesses or its overall impact on the island. The council deadlocked three to three on the motion to grant the mini-golf permit. The mayor cast the deciding vote, in favor of the project. After the vote, a frustrated council member asked if the city could not regulate the town's look and feel, what was the point of building permits?[40]

In 1975, a new mayor, Margaret Doud, was elected. Doud was clear about her priorities, saying the most pressing problem in the community was city planning. Her concern was that the business district was moving away from the island's historic architectural style. "I don't think we can let the town deteriorate, architecturally, very much longer."[41] Doud knew a tough road lay ahead of her. When W. Stewart Woodfill had been chair of the Mackinac Island State Park Commission, he too had lamented the appearance of the town. Woodfill was not a man to shy away from a fight, but he decided the opposition from the island's business community to new zoning regulations would be such that he chose not to propose any changes to the law.[42] Years later, with a clear demonstration regarding the profitability of historic tourism found in the paid attendance to Fort Mackinac, Doud chose to fight that fight. After she was elected, she quickly announced the appointment of a planning committee with the objective of drafting new legislation.[43] By the end of 1975, the committee had produced a proposal that would organize and control construction on the island, and which included a historic preservation component.[44]

There was stiff opposition to Doud's agenda. The local newspaper preferred voluntary action, while lauding the work to maintain the "character of architecture in its classic Victorian form" and agreeing that doing so had "long-term economic considerations, mainly the preservation of the unique attractiveness of the island—its historical atmosphere." As for the mini-golf controversy, the paper dismissed the fight as nothing more than finding a use for a vacant lot behind a local establishment. The paper believed that, in the future, a process that embraced "quality, beauty, courtesy and friendliness" was preferable to "some bureaucratic enforcement of a rule."[45]

Doud, who had been a member of council during the mini-golf course discussion, was skeptical of the "courtesy and friendliness" the city could expect from a business owner who wanted to build a profitable, if tawdry, enterprise. In the place of governmental courtesy and friendliness, she took every opportunity to ensure the "bureaucratic enforcement of a rule" that was on the books while also seeking to enact new laws. In order to do so, the planning committee was given additional authority to review all building plans, including several long-delayed projects, going back as far as 1966.[46] Doud was able to pass an ordinance establishing a city building department. New rules required the involvement of a licensed architect or professional engineer in any residence larger than 3,500 square feet and all commercial or public assembly buildings.[47]

Doud also acted to vigorously enforce any ordinance she could find that might impact the city's appearance. Her initial tool was a forgotten 1952 law that required barns to be located at least one hundred feet from residences. As it turned out, some buildings in the town's business district appeared to be in violation of this law.[48] Doud made it clear that in the future, there would be strict enforcement of all existing local and state rules.[49] Her message was that those who preferred a laissez-faire approach to the town's appearance, or who might try to stonewall her planning committee, would be wiser to work with her and her committee to draft new laws rather than face the unpredictable possibility that there were other forgotten but enforceable laws such as the barn ordinance already on the books.

The state park staff quietly supported Doud's endeavor. In 1974, an exhibit opened in the commission's Island Visitor Center about the island's architecture. The exhibit ended by describing the architecture of the "tourist era." The exhibit's closing statement was an implicit endorsement of Doud's 1975 campaign platform. Lamenting that buildings put up in the 1930s and later had "little regard for our understanding of what makes Mackinac Island special," it was noted that they "greatly diminished Mackinac Island's unique character." The island faced the choice of maintaining the "resort era quality of Mackinac Island" or to "continue to let future construction take its course."[50]

Doud's vision gained an additional ally: Hollywood. In February 1975, stories began to appear that Hollywood was "sniffing around" both the island and the Grand Hotel to use them as the setting of a new feature film.[51] R. D. "Dan" Musser, president of the Grand Hotel, had laid out the red carpet for Hollywood representatives, in the hopes of convincing them to make a film on the island.[52] Musser knew, as had Woodfill before him, that a movie would bring invaluable publicity. It would, but its more lasting impact was on the island's appearance.

In the summer of 1979, the film *Somewhere in Time* was shot on the island. It was a period romance set in the Victorian era. The movie was released in 1980. Not everyone liked the movie or wrote kindly about the setting. One newspaper reporter said, "Floating out there somewhere between the ozone of ostentatiousness and the time warp of old-world wealth, the Grand is still, well, grand."[53] Perhaps it was ostentatious, but the film's fans loved the time warp portrayed in the film. As another newspaper review of the movie put it, "The movie was shot almost entirely at the neo-classical Grand Hotel on Michigan's Mackinac Island and scenes of the hotels and its spectacular grounds are worth a week's daydreams in themselves. The hotel is a central character in the story." The same review concluded that one of the movie's best scenes "describe[s] to the audience the Victorian dreams in the hearts of young women."[54] Although Mayor Doud could not assure the preservation of a young woman's Victorian dreams, she worked very hard to ensure that potential tourists who had such dreams could enjoy them in the island's carefully preserved Victorian architecture. With the commission, and Hollywood, having her back, with money flowing into the tills of local merchants, and through using considerable personal skill, Doud accomplished her goal. It would take years to see the project to its successful conclusion, but by 1981, the necessary zoning laws were in place.[55]

The new laws, did, on occasion, create controversy within the business community. For example, in 1988, Pat and Alice Pulte, downstate developers who had purchased the Chateau Beaumont Hotel, were denied a building permit. They complained, alleging that the city architect, who was also a practicing architect, had a conflict of interest. They had chosen not to hire him for their project, and in return, they alleged, he had held up their project. The Pultes further claimed that the architect did this routinely. If you did not hire the man for a project your permit would be held up. In the case of Chateau Beaumont, the architect had asked for more information about paint. Pulte fumed, "The problem here is conflict of interest, not colors." The architect, however, who estimated that he spent 25–30 percent of his time on his official duties, responded that the real problem was the small pool of architects to choose from in northern Michigan and the unfamiliarity of most of them with the intricacies of Victorian style. Because of this, the initial submission for many projects were not fully compliant with the Victorian appearance required by city code. Mayor Doud summed up the problem best when she said, "In a small community you're bound to have some conflict of interest, because some people wear more than one hat."[56] In the end, the Chateau Beaumont project seems to have been approved, and the issue faded away.

On occasion, the city, including Mayor Doud, could look the other way, favoring economic development over architectural purity. In 1980, for example, a developer who had purchased one of the largest privately owned pieces of land on the island announced plans to subdivide a portion of it into thirty-eight one-acre parcels for single-family homes. Critics railed against the idea. The *Detroit Free Press* editorialized that it was "the worst threat to the Island since the British hauled a cannon up to the heights overlooking the fort in the War of 1812." The project, in the *Detroit Free Press*'s opinion would "suburbanize the gem of the straits."[57] The city's government, however, saw it otherwise. It unanimously approved the project.[58] Of the Detroit paper's editorial, Mayor Doud said, "It's been controversial and I've held that the *Free Press* coverage has been rather slanted."[59]

If one of Doud's pressing issues was preserving local architecture, the other was the ongoing problem of paying for the infrastructure needed to accommodate over a million visitors a year. With a permanent population of only about eight hundred people, the idea that local property taxes could support the various public infrastructure needs created by the visitors, issues like bathrooms and street maintenance, was challenging. The island government often sought special state aid so that projects like a sewage treatment plant that met contemporary pollution standards could be constructed,[60] but as the park commissioners had learned, the city found that the state legislature was often tightfisted.

Everyone knew more local money was needed, but as often happens when the question of taxation arises, everyone wanted someone else to pay the tax. One solution was to expand the tax base by obtaining revenue from the visitors who came to the island, collecting it as a surcharge on ferry boat tickets. It was not a new idea. The Park and Harbor Commission of the early 1940s had come to the same solution. In 1977, the city adopted an ordinance establishing a 1.5 percent tax on the gross revenues of the ferry boat companies serving the island.[61] As she had with the issue of architecture, Mayor Doud led the charge. As the local paper, which supported the tax, wrote, "Margaret Doud deserves recognition by all parties for having had the foresight, courage, and persistence to do something practical about the problems."[62]

The ferry companies did not share the newspaper's positive opinion regarding Mayor Doud's vision. They declared the tax unfair and unconstitutional and fought it in the courts.[63] Bill Shepler, leader of one of the ferry companies, acknowledged Doud's broader point, when he said the problem was not that the tax wasn't needed, but complained her answer was a poor one. The tax should have been assessed more equitably, he believed, including a tax on hotels by the room, restaurants by the seat, and bars by the stool.[64] Shepler had a point; island hotels, restaurants, and bars were all also profiting from visitors but were not being asked to collect money from them for the city, nevertheless a passenger tax collected by the ferry boat lines was a logical way to make every visitor pay some of the costs the community bore because of their presence. The court cases went on for years, including appeals to the US Supreme Court, with the ferry lines consistently losing. When the appeal process regarding the legality of the tax was exhausted, the ferry lines fought in court over how the money would be paid to the city. Finally, in 1984, the matter was resolved through a settlement that applied the tax in a way that offered some important concessions to the ferry companies.[65]

Between them, Margaret Doud and Eugene Petersen had secured historic tourism as the principal attraction on Mackinac Island. Petersen had solidified the beginnings made by Woodfill and Governor Williams while Doud had led a sometimes reluctant community to both preserve the community's Victorian architecture and find a way to resolve the chronic problems caused by the crush of tourists visiting the small community. It was a singular accomplishment.

CHAPTER 8

The End of an Era, 1961–1990

The emphasis on historic tourism that swept over the state park and had a significant impact on the workings of the Island's municipal government was felt less fully by the island's business community. This was perhaps to be expected, as the swell of tourists buoyed the local tourist economy, making change less necessary. There were members of the business community who not only embraced change but were the agents of that change. Others were grudging about new trends or preferred to rely on the old ways of business and were critical of those who benefited from change. The passage of time also led to the most profound change, the passing of the leaders who had guided the island for from twenty to fifty years.

Marvin May, who had purchased the Murdick fudge shop on the island in 1940 after working in it for about a decade, was a merchant who did not like many of the changes and did not like the new businesspeople making the changes. Of his post–World War II competitors, May complained, "They didn't come here to make it [fudge] famous, they came here because it was famous, and they could cut in on it."[1] May's words were largely a matter of opinion. But others did more than simply criticize their competition, and in doing so continued an unsavory aspect of island business dating back decades.

Island merchants had been criticized for more than a century for their sharp dealings with customers and, less obvious to the public, equally sharp practices regarding employees. Customer practices could be so poor that some merchants were called "human sharks" in the nineteenth century.[2] In the 1920s. carriage operators had become so unruly, storming aboard passenger ships trying to solicit business, that ship officers banned them from coming aboard.[3] The freewheeling carriage operators also had a long and unhappy relationship with the park commission and some island business people, including W. Stewart Woodfill.[4]

Employee relations were another area where island business practices had sometimes been questionable. Woodfill's behavior at the Grand Hotel in the 1930s when he refused to pay entertainers stands as a particular example of worker exploitation.[5] In the 1960s, when a statewide minimum wage was proposed, the Mackinac Island business community opposed the idea. The feeling was wages were a matter to be decided between employer and employee, a system that more often favored the employer. W. Stewart Woodfill, owner of the Grand Hotel, claimed that since employees received tips and free room and board, there was no need for a minimum wage. Woodfill implied a minimum wage would hurt his business, claiming, "There are no prosperous summer hotels."[6] This was a disingenuous statement from a man whose hotel had been regularly turning a profit since the 1940s.[7] If there had to be a minimum wage, W. Stewart Woodfill asked that hotels that operated less than six months a year—that is, hotels like his—be exempt. Bill Doyle, on behalf of the carriage operators, also opposed the minimum wage bill, repeating Woodfill's point regarding the tips drivers received.[8] The foot dragging on this issue by island businesses garnered enough attention that the state American Federation of Labor-Congress of Industrial Organization (AFL-CIO), which favored the idea, complained about it.[9]

A decade later, in 1979, the Michigan Department of Labor alleged "a group of Mackinac Island businessmen" was conducting a "massive violation of state labor laws" that involved many employees who collectively had been wrongfully denied $25,000. Among the allegations were violations of the state's minimum wage laws, forced overtime without overtime pay, required "employment fees," and required "work bonds" (money taken from a paycheck that will be given back to an employee if they stay until the end of the season) along with allegations that employees were forced to live in housing owned by their employer, with the rent money automatically deducted from their pay.[10]

When the allegations became public knowledge, some of the business owners involved admitted no wrongdoing, asserting instead the employees in question should be thankful to them. The owner of one of the restaurants the State had accused of wrongdoing claimed that "his employees were fortunate to be able to live in rooms he supplied at the cost of $25 a week." He added, "Mackinac Island attracts many of the upper middle income children [for summer jobs] and these kids are sexed up, boozed up and drugged up. I think the parents should pay us for taking care of them." The matter was quietly resolved, most likely because state law said that if restitution was voluntarily paid to the workers whose financial rights had been violated no criminal charges could be filed against the employer.[11]

In 1989, there occurred a far more serious crisis regarding island employees than the failure to follow labor laws. A fire in September 1989 swept through a building that housed summer workers, killing two summer employees.[12] Widespread newspaper reports revealed that the city had no ordinance requiring the regular inspection of employee housing.[13] As one reporter wrote, "If Mackinac Island is the place time forgot, it now turns out it's also the place fire inspectors forgot."[14]

State fire inspectors were soon on the scene, both investigating the tragedy and inspecting all employee housing on the island. By the end of October 1989, they had inspected sixty-nine buildings that housed approximately two thousand summer workers. Of these, six were in full compliance, two were closed as fundamentally unsafe, and the balance were cited for five hundred fifty-seven code violations in total.[15] The island's business community responded to the tragedy with a genuine effort to fix things. Over a million dollars was spent on renovations and repairs during the winter of 1989–1990. In the spring of 1990, the city hired its first full-time building inspector.[16] One of the state fire marshals who had performed the initial inspection in the fall of 1989 commented in the spring of 1990, "Frankly, we were skeptical that they could get the work done by this spring. But in my 25 years in the regulatory business, I have never seen people respond like they have on the island. They have busted their butts."[17]

While most people accepted responsibility and worked very hard to fix building code violations, some continued to speak in a voice that reflected the island's unhappy history regarding labor relations. The owners of the building that burned, who would eventually be criminally charged, perhaps understandably placed the blame for the fire on fate. One of the two men who owned the burned building said, "We're not bad people. Yes there was a tragedy. But tragedies happen." Pointing to the community's work to make things right, he added, "Everybody is making a very valiant effort to make things safe."[18] Others, however, went beyond saying bad things happen to good people. Some claimed the state's attorney general was using the fire "as an excuse to grandstand." Others grumbled that the deaths had been caused by "the fire, not code violations."[19] And some blamed the victims. Summer employees they said were "rowdy summer workers [who] take batteries from smoke detectors" and "pile trash in their rooms and break windows."[20]

The tragedy saw public relations come into play. To address the island's "tarnished" image, a public relations firm was hired jointly by the city, the Chamber of Commerce, and the Mackinac Island State Park Commission.[21] Perhaps the outside firm was unneeded. The island's tourist industry continued its long tradition of employing master publicists to spin a positive image. One of the most notable yarn tellers of the 1970s and 1980s was Wilmer T. Rabe, who was associated with the Grand Hotel.

Rabe could enrapture almost any reporter with his imaginative creations. One of his long-lived inventions was the "Mackinac Island Stone Skipping and Gerplunking Club." The group held an annual stone skipping contest on July 4. Throwing stones into the lake was old hat, but the way Rabe presented it, the story was in every newspaper. The event began with the "Lord High Commissioner of the Solemn Sport of Stone Skipping," intoning the words, "Let he who is without Frisbee cast the first stone." This admonition was then followed by one spoken by event's field judge, who proclaimed, "Gentlemen, start your stones." What the contest winner received varied each year, but in 1981, the winner received the

Little David Trophy, a somewhat battered silver cup left over from a 1967 bicycle race, a seventy-five-pound stone, and fifty pounds of fudge.[22]

Nothing stirs up public interest like controversy, and in 1976, Rabe manufactured several. The first was the story of two stone types, "olivine gabbro" and the "pretty pink ones" that were used by contestants despite allegedly having federal protection. Rabe conceded, however that this violation of federal law paled in comparison to the "Incident at Lock 123." That was a matter so horrible Rabe refused to release any details or discuss it in public, a time-honored way to get reporters to keep asking about it.[23]

The next year, Rabe expressed the deep concern that past competitions had depleted the island's supply of suitable rocks. He asked "stone lovers round the country" to fill the gap by sending a stone and a self-addressed, stamped envelope, promising that an official scorecard would be mailed to the rock donor, detailing how the stone fared in the tournament. About two weeks later Rabe, doubled down on the stone shortage controversy by reporting competitors wanted to use "perfect" commercially produced stones, but it was unclear if these were acceptable. The high commissioner was to rule on the matter.[24] Apparently the high commissioner did allow the use of commercially made stones, but only if they had first been boiled by the tournament veterinarian, undoubtedly to avoid infecting the local rocks.[25]

Rabe not only made-up great gags, he could also turn legitimate news into an irresistible yarn. One example was the 1976 discovery that Russian citizens were banned from Mackinac Island. The actual explanation was that this was a byproduct of the Cold War. The United States and the Soviet Union had both instituted bans forbidding citizens of the other nation from visiting certain parts of their country. Originally, the bans applied to places of military significance, but the practice eventually devolved into an endless cycle of retaliatory bans, with one nation banning travel to some site in their nation, and the other evening the score by establishing a new travel ban of their own. At some undetermined point, Mackinac Island had been declared off-limits to Russians. The factual explanation, however, was boring. Rabe improved on it substantially. According to him, the Russians were banned from the island because the Soviets wanted photographs of the deadly cannons found within Fort Mackinac. Rabe further made the shocking discovery that the Mackinac Island State Park Commission was compromising national security. If the Russian spies forgot their cameras, they could nevertheless complete their mission by buying a postcard in the fort gift shop with a very nice picture of the guns.[26]

Publicity continued to be a mainstay of the island's tourist trade, but a new breed of business owner, exemplified by Harry Ryba, did not worry as much about it as had his predecessors. Even the terrible publicity received after the tragic fire of 1989 did not seem to bother Ryba. His assessment of the matter was blunt: "People will come to Mackinac Island come hell or high water."[27] Ryba was a man who did not much worry about what the papers, or, for that matter, what others on Mackinac Island, had to say about the island or his business. That was a good thing since he was one of the principal figures changing the island's business community. In the process, he made many enemies.

In 1930s, Harry Ryba was working as an employee for a Detroit creamery.[28] He was smart and hard-working, and he saved his money. With some of his savings, he bought a candy shop, which his wife ran. During World War II, Ryba found work in one of the many wartime factories in the Detroit area. Factory work, however, was not for him, and soon after the war ended, he quit the factory and began working full-time in the candy store.[29]

In 1946, the store began to sell fudge made by Francis Murdick, a grandson of Rome Murdick, the man who had first made fudge on Mackinac Island. Francis rented space in Ryba's store and made his product using his grandpa's recipe. The confection sold well. Harry eventually bought out Francis, including grandpa's fudge recipe. Ryba took the fudge on the road. In 1949, over the ten-day run of the Michigan State fair, which was then held in Detroit, he sold three thousand pounds of fudge at a dollar per pound. That was more money than he had ever seen. He had found his calling.[30] In 1959, Ryba decided the best place to sell fudge was Mackinac Island so he relocated to the island and opened a fudge shop.[31]

Ryba quickly began to shake up how island business was done. He claimed he was the first to make fudge in a store window.[32] Ryba added to total sales of his window-made fudge with the slogan, "When you see it made you know its fresh."[33] His son-in-law added suspense to Ryba's shop window. In making the confection, molten fudge is poured onto a marble table to cool. As the cooling takes place an employee works the thick liquid with a paddle, pushing it away from the table's edge and eventually shaping it into a brick. Ryba's son-in-law developed a routine in which he chatted up his audience while seemingly paying no attention to the molten chocolate. As the patter continued, the audience would draw closer, sure the fudge was about to pour over the table's side. But at the last possible second, a swift move with the paddle saved the day, or, more accurately, the fudge. And if appealing to the visitors' eyesight and employing a bit of melodrama was not enough, Ryba also went after their nose by relocating his store's exhaust fans from the back of the store to the front, blowing the smell of freshly made fudge into the face of every passing tourist.[34]

Ryba not only made his fudge with flair, he marketed it the same way. His competitors put fudge in plain white boxes. As early as 1964, Ryba opted for eye-catching pink boxes, carried home in pink bags.[35] When the wife of President Lyndon Johnson visited the island, Ryba knew he couldn't get close enough to her to give her a box of fudge. W. Stewart Woodfill was personally leading her walking tour around the Island and Woodfill was not about to let some candy man shoehorn his way next to Lady Bird Johnson. So Ryba came up with an alternate plan. He "persuaded" a person who was working in the First Lady's overnight accommodations to place a box of his fudge on her dressing table. When a gracious thank you note was received from the White House, Ryba wasted not a minute in having it framed and prominently hung in his store.[36]

Ryba tried to get the carriage tour operators to promote his product. When he learned that there was a policy against their making endorsements, he rephrased his request. "I told them that if they were asked which fudge was best to just reply that 'I don't know, but I eat Ryba's myself.'" And to help make sure that was true, he offered a permanent discount to carriage operators or anyone else living on the island.[37]

Although many of Ryba's activities came down to clever marketing, when he felt the need, he used a sharp elbow to get his way, or at least slow down something for which he didn't care. For example, Ryba was sued by May's Fudge for using May's copyrighted phrase "Mackinac Island fudge." The violation was obvious, and Ryba's court case futile to the point of being frivolous, but it allowed Ryba to keep using the phrase for as long as possible. Perhaps Ryba fought the fight to decide on a new slogan. When the case ended, the confection Ryba sold, formerly advertised as "Mackinac Island fudge," became "The best on Mackinac Island."[38]

Ryba not only ran a fudge shop, he bought restaurants and several hundred rental bicycle licenses. He also bought hotels on the island, including the Lakeview and the long-troubled Island House, offering visitors a less expensive way to extend their stay and spend the night. "Most people can't afford a $22 a night room, for example. They're looking for something at $6, $8, or $10."[39] Eventually Ryba owned his own ferry boats, which sailed from his dock in Mackinaw City, which offered free parking.[40]

Ryba may have been among the first to realize that there were two tourist markets on Mackinac Island. One consisted of the people W. Stewart Woodfill housed at the Grand Hotel. They, however, had become the minority of visitors. In 1961, the Chamber of Commerce estimated that most visitors spent only two or three hours on the island.[41] In 1966, a study reported more than 400,000 visitors came to the island, of which 376,700 were day tourists.[42]

What was happening was the fruition of a fundamental shift in who could enjoy a vacation. The automobile had made it possible for working class people to take a vacation. The possibility of them doing so greatly expanded when, between 1930 and 1940, the majority of America's industrial workers were granted a paid vacation.[43] Particularly in Michigan, with its well-paid automobile factory employees, driving up to Mackinaw City and taking the ferry to the island for the day was an outing well within the family budget, even if the budget still called for spending the night in a camper or a modestly priced Mackinaw City hotel.

Ryba may have invented a new way to exploit the island's tourist industry, but being successful did not make him well-liked. Islanders developed strong opinions about Ryba. W. Stewart Woodfill's opinion was among the more complex. Woodfill somehow blamed Ryba for the day visitors he disliked, saying Ryba was attracting tourists who came "with their passion for Island-made fudge and Hong Kong souvenirs."[44] Not only did Woodfill object to their taste in confections and gifts, he also was appalled by the way these new arrivals dressed. "At luncheon it's a horror! These days the tourists, our own guests, will come in any form of deshabille. Hot pants! God knows, hardly any pants at all! Rags! An absolute disgrace."[45] If Woodfill thought the people Ryba built his success on were disgraceful, Woodfill was amazed by Ryba's financial acumen.[46] Ryba knew his way around a ledger at least as well as the master of the Grand Hotel. Others were less nuanced in their criticisms. Among the more predictable complaints was that Ryba's fudge "tasted terrible."[47] Some called him a "candy carpetbagger." Some attacked his character. "He's a nickel grabber and that's it." Another member of the island business community summed up Ryba this way: "There are people who to get something will step on others to get it, and he's that kind of person." A fellow businessman who had known Ryba for twenty years, usually thought of him as an amiable rascal. But he added that, on occasion, Ryba could be "a miserable little bastard."[48]

Ryba knew people didn't like him and he didn't care. In fact, he often did not like them. Ryba said, "A lot of people here don't like me. Whatever was up for sale, I bought it. I grabbed it." Reflecting his different vision for the island tourist business, he added, "When I got here these people, they had no vision."[49] Ryba also had a few words in response to Woodfill's lament about the "deshabille" of island visitors: "Some people want to keep the Island for the 'high class' but today it's the 'Fudgies' [day tourists] that are coming over in streams."[50]

Ryba focused on the reality of the tourist trade. Part of that reality, he realized—as had Woodfill when he marketed the image of the Grand Hotel, Petersen when he sold the island's history as director of the Mackinac Island State Park, and Mayor Doud when she enshrined the island's Victorian architecture in zoning ordinances—was that no matter how long a tourist stayed, what brought many visitors was the island's appearance. Victorian charm had come to supplement, and in many ways replace, the island's natural beauty as the reason why visitors came.

As Ryba expanded his businesses to include hotels, he demonstrated his understanding of the need to preserve the island's architectural past through his restoration of the Island House Hotel. Opened in 1852, the building had a complicated history, eventually becoming government owned in 1938.[51] In 1942, the Mackinac Island State Park Commission, which had control of the building, leased it to a religious organization that used it as its headquarters.[52] After many years and as many disagreements, in 1948, the park commission (but also involving the city, other state agencies, and the state legislature) ordered the religious organization out of the building.[53]

The park commission then rented the building to a private organization which reopened it as a hotel.[54] By 1969, years of deferred maintenance had left Island House derelict and condemned by the fire marshal.[55] Most people believed the building, one which in its day had been a leading example of island architecture, would have to be torn down.[56] Harry Ryba disagreed. He obtained a long-term lease for the hotel from the commission and restored it to its previous glory. The project, which was originally estimated to cost $250,000, cost Ryba over one million dollars.[57] When the restoration was finished, Island House's storied past and historically sensitive restoration was acknowledged through a Michigan historical marker.[58] Ryba materially contributed to the island's Victorian look and feel, something Ryba knew mattered, but Ryba embraced historic architecture with a clear eye regarding profit. Some described the restoration of Island House as a labor of love, but Ryba's love for the building was considerably helped by the commission attaching 235 bicycle rental licenses to the hotel and the assigning its sole liquor license to Island House.[59] Reconstructing the building mixed together sound history and good business.[60]

Beyond what Harry Ryba did at Island House, Mackinac Island's hotel business was also impacted by two other major changes: the opening of a new large hotel eventually named Mission Point Resort and the reconstruction of the Grand Hotel. Mission Point Resort had a complicated history that began without a connection to tourism. It was built by a religious organization named Moral Re-Armament (MRA). When MRA had first come to the island, the Mackinac Island State Park Commission had given the organization the use of Island House.[61] In 1948, MRA left Island House with considerable hard feelings on the part of both the MRA and the commission.[62] Eventually, MRA purchased a fifteen-acre tract on the "Mission End" of the island, across town from the Grand. In 1956, MRA opened a million-dollar complex. Eventually the MRA property included a theater, large housing complex, a kitchen and dining room that could serve one thousand, and film and television production facilities, all of which were added to pre-existing buildings also owned by MRA on the east side of the island.[63]

MRA's objective was reviving the moral character of the world through the application of Christian principles, as defined by its founder, Rev. Frank Buchman.[64] Buchman and the MRA were controversial among islanders. Some supported the movement as a steady source of year-round income with the added benefit of drawing organization members to the island during tourist season.[65] Others called it a religious cult whose presence dissuaded nonbelievers from visiting the island.[66] Mackinac Islanders continued to argue about MRA, but the argument came to an end when MRA decided to relocate its headquarters.[67] In 1965, the organization's fifteen-acre island complex was deeded over to a private liberal arts college, named Mackinac College.[68] The college built additional buildings on the site. Very quickly however, the college found the problems that had plagued MRA's operation, remoteness and limited access during the winter, meant it could not attract enough students to survive. In 1969, the college was put up for sale.[69] It was purchased by Rev. Rex Humbard, whose immensely popular television ministry, the "Cathedral of Tomorrow," was then seen on more than two hundred television stations and created a substantial financial base.[70] Humbard opened a seminary in the abandoned college but added to it a vacation opportunity for his followers, promising "Vacations with Inspiration." Those who came availed themselves of "worship, communion, study, and recreation."[71] Humbard's ministry collapsed in 1973 when federal and state charges were filed regarding the church's financial dealings.[72]

In 1973, the property was again up for sale. The seminary was closed but the resort operation was continued, leased by a third-party operator.[73] The property would change hands several times before it was fully renovated from a religious center and college into a successful hotel that was named Mission Point Resort, the second largest on the island, with approximately 250 guest rooms. At the time, the Grand Hotel had approximately 350 guest rooms.[74]

In comparison to the newly built Mission Point Resort, the Grand Hotel seemed increasingly dated. As one person wrote, "It [the Grand Hotel] had faded badly over the years and big spenders were taking their dollars elsewhere."[75] The hotel's interior had last been seriously updated just after World War II. What guests experienced at the Grand in 1970 was a room furnished from the Sears Roebuck catalog and a large number of chairs upholstered in plastic that were easily washable and, to the eye of the hotel owner, W. Stewart Woodfill, looked "just like leather." The wife of Woodfill's nephew and Grand Hotel President R. D. "Dan" Musser, Amelia Musser, had been an art major in college and could see the problems with the Grand Hotel's interior. She saw the need to transform the Grand Hotel into what she hoped would be a "heartwarming showcase."[76] In 1977, the Mussers decided the time had come to revamp the interior. The Grand's problems, however, went far beyond its interior decorating scheme. Since 1966, it had been known that the hotel's east end was tipping forward, caused by the rotting of the wooden foundation upon which the structure rested. The problem was ignored for as long as possible, but in 1976, the year before the redecorating campaign, the hotel began a multi-million-dollar structural project.

The management of both the Grand Hotel and Mission Point Resort joined arms with Harry Ryba in working to preserve the look and feel of the island's architecture. They all worked closely with the City Planning Committee and other appropriate city agencies to ensure a harmonious appearance.[77] By 1990, these leaders of the business community, working with the municipal government and the Mackinac Island State Park Commission had transformed the island into a Victorian showpiece. But the people who had been responsible for this accomplishment were passing from the scene.

G. Mennen Williams, who had embraced the idea of a historically centered Mackinac Island and, to accomplish this idea, had transformed the park commission, died in 1988. His deep love for the island was demonstrated by his decision to have his body buried there.[78] In 1985, Williams's longtime nemesis, Bill Doyle, attended his last meeting as a member of the Mackinac Island State Park Commission. Doyle died in 1988.[79] In 1990, Williams's one-time right-hand man on the commission, James P. Dunnigan, was writing his reminisces.[80] He would die in 1996.[81] The commission had lost most of its longtime members. Eugene Petersen, another of the individuals who had reshaped the island, noted, when he retired in 1985, that with one exception, the long-served commissioners with twenty or even thirty years of experience had been replaced by people with less than six years of experience.[82]

Williams was gone and the commission was made up of new people. The park staff was in new hands. The business community was also changing. In 1990, Harry Ryba was in his early eighties and, after thirty years of whirlwind activity, slowing down.[83] He would die in 1996.[84] Lesser lights within the business community also passed from the scene. Wilmer T. Rabe, who invented the solemn sport of stone skipping and so many other ploys to garner publicity for the island, retired in 1987.[85] All these transitions were foreshadowed, however, when the grandest of the island's grand men, W. Stewart Woodfill, left Mackinac Island.

As the end drew near, Woodfill seemed to understand what must come. In 1975, when President Gerald R. Ford came the island for an overnight visit, Dan Musser, then the president of the Grand Hotel, and not W. Stewart Woodfill, the hotel's owner, greeted the first US president to visit the island while in office. Woodfill explained his taking a back seat on such a momentous occasion by saying, "Sooner or later you have to let the younger people take over."[86] It was a gracious public statement, perhaps the last gracious act, public or private, that Woodfill would perform in what became an increasingly ugly transition at the Grand Hotel.

Woodfill's personality likely made his final days at the Grand Hotel inevitably troubled. As Eugene Petersen wrote,

> the transfer of power in a dictatorship is a contradiction. It's doubtful if anyone could have run the hotel to his [Woodfill's] complete satisfaction, yet time had passed him by. He preferred the façade over the foundation. He insisted on impeccable dining room service by subservient male waiters who served properly attired guests. But the hotel management had to deal with minimum wage, affirmative action, sexual equality, youthful clientele in casual dress, and a building that rested on wooden stumps.[87]

This would be a transition, Petersen accurately thought, that Woodfill was incapable of making gracefully.[88]

Dan Musser, Woodfill's nephew, was heir to the throne. Woodfill had no children. Dan Musser's father had died while he was young and his mother remained a widow, with modest resources. To help his sister, Woodfill paid

Musser's college bills.[89] While still in college, in 1951, Musser began working at the Grand Hotel.[90] As time passed, it became clear that he would take over. Woodfill's intentions were signaled in 1958, when he made Musser a manager within the Grand. Some other managers opposed making someone so young and relatively inexperienced a manager. Woodfill's response was to fire the ringleader of the critics.[91] When Woodfill became chair of the park commission, he dedicated a large portion of his time to that job, trusting Musser to run the hotel. In 1960, Musser was named the hotel's president. He took full charge of the operation in 1962.[92]

In name, Musser was in charge, but Woodfill kept his eye on everything and often meddled in day-to-day operations. He himself once admitted, "I drive the kid crazy. At least 10 or 15 times a day I call up there to see what Musser is up to."[93] Musser's patience and tact with his increasingly irascible uncle made the relationship work. But slowly, and then with increasing speed, things unraveled. What brought matters to a head was the need for change and the money it would cost to do it. By the mid-1970s the hotel faced new competition on the island at a time when the Grand Hotel interior was increasingly dated and major structural work on the building was necessary. Woodfill was unwilling to address any of this. Mussser recalled one conversation when, as he phrased it, "Mr. Woodfill and I got into one hell of a fight on parlor furniture."[94] Despite that fight, and likely other ones, the interior renovations went forward.[95] The real disagreement, however, was not about interior design but the hotel's structure, literally the building's foundations which were made of wood and rotting. Woodfill feared that addressing the Grand Hotel's basic structural problems would open a can of worms. Matters were not helped when the preliminary surveys for the project led to headlines about the exaggerated length of the hotel's much praised porch, something that must have greatly annoyed Woodfill, who was the perpetrator of the fraudulent story about the porch's length.[96]

Things were unraveling, but true to the tradition of Grand Hotel, the public was the last to know. According to a newspaper story published in 1978, years earlier, Musser and Woodfill had been standing in the hotel's lobby discussing the season just ended when Woodfill supposedly said, "This place looks dingy—we have got to do something about it." As for the structural repairs, the paper reported that Woodfill was completely supportive, telling Musser to "do the right thing and get it done." The newspaper added, "Dan Musser credits Woodfill with the perceptive order to reconstruct and do whatever it takes to do the job right."[97] It was a good story, but like so many of the Grand Hotels stories, it was not true. Woodfill did not want any of the changes and was beside himself at the amount of money Musser was spending.[98]

In 1978, Woodfill was done. He dictated a bill of sale that he had presented to Musser. Woodfill, despite offering to sell the hotel to Musser, may have not wanted the deal to occur. He seemed to feel that what he proposed, in particular the amount of money he wanted, would delay his nephew, or at least require a protracted negotiation between the two men. Woodfill seemed oblivious to the strong relations his nephew had built with several lending agencies. Musser accepted the bill of sale as written and quickly found the financing needed to make the purchase.[99]

With nothing needed to complete the sale but a few signatures, Woodfill invented and released his last great hoax: a story that he was giving the hotel to Musser. The announcement was shared as an exclusive with the St. Ignace newspaper. Woodfill gave his nephew no warning about it. Instead, he informed, really demanded of the newspaper's editor, that "Mr. Mussser be apprised of his [Woodfill's] gift through this announcement."[100] When he read the story, Musser knew it was false. His uncle was not giving him the hotel. But, on the advice of his attorney, Musser remained silent. The sale of the Grand Hotel was finalized in 1979.[101] Woodfill never explained why he originated the false newspaper story and Musser never offered a public explanation for his uncle's behavior. Woodfill's biographer guessed that it was a final prank by a man who had been a prankster all his life, but that guess seemed unconvincing even to the biographer. No one, Woodfill's biographer wrote, really knew what was going on in Woodfill's mind.[102]

On September 19, 1978, Woodfill took his final leave from the island.[103] He left without any public notice that it was his last day, but knowing he was never coming back.[104] He was accompanied on the trip only by Dan Musser. The two men took the ferry to Mackinaw City, and from there they went to the airport in Pellston, Michigan, where Woodfill boarded a plane for a trip to Arizona.[105] Woodfill died in Arizona on March 4, 1984.[106]

What Woodfill was thinking on his last ferry ride from Mackinac Island will never be known. Perhaps it was a mix of fact and fiction, some of it fiction he himself had created. But leaving the island with a mix of fact and fiction was a tradition going back to Harriet Martineau. In 1837, her mix of the two had begun the making of Mackinac Island as a tourist destination. In 1837, as in 1990, Mackinac Island was a place where those who lived or did business there experienced one reality but often portrayed the reality of the island to visitors in a different way. The story of the place created by the island's tourist industry and shared with visitors is a wonderful tale. The only thing more fascinating than that mix of fact and fiction is the how the island's tourist industry really came to be.

PART 2

The Shoreline

CHAPTER 9

The Beginning, 1895–1932

Tourism on north and south shores of the Straits of Mackinac began in the late nineteenth century when the railroad arrived in Mackinaw City and St. Ignace. However, for many years, most of the tourists who came were bound for Mackinac Island. They got off the train and boarded a waiting ferry to take them to Mackinac Island. If they stayed in either Mackinaw City or St. Ignace, they found two relatively small communities with surprising histories.

Although Europeans lived on both sides of the Straits of Mackinac in the seventeenth and eighteenth centuries, eventually, Mackinac Island's small village became the area's principal European community. St. Ignace was the first place on the straits inhabited by Europeans. The first Europeans to live there were French missionaries, who arrived in 1671. About ten years later, the French army established an outpost, named Fort de Baude, near the Catholic mission. Both the priests and the soldiers abandoned their sites on the straits' north shore. The French army left in 1701 and returned to the straits in approximately 1715, constructing

Fort Michilimackinac on the south shore of the straits. The missionaries left St. Ignace in 1705. Throughout the eighteenth century the site was mostly uninhabited by Europeans. A new American community began to grow at the beginning of the nineteenth century. In 1814 a naval officer commented that five or six families were present. In 1833 a Protestant missionary reported that the community numbered 180.[1]

In 1874 St. Ignace had 326 residents.[2] The town was on the cusp of a lumber-fueled economic boom.[3] The arrival of the railroad in 1882 made St. Ignace the southern terminus of Upper Peninsula railroad lines, and the community's railroad was soon linked by a ferry boat to the railroad tracks in the Lower Peninsula that terminated in Mackinaw City.[4] The local economy was rapidly growing, and boosters claimed St. Ignace would eventually become as large as Detroit or Chicago.[5]

When St. Ignace's leaders thought about what would happen if the trees ever ran out, what they saw was a farming town. As one newspaper put it, "As soon as farm development comes St. Ignace will take her merited place . . . in the Upper Peninsula."[6] The belief that agriculture would come to St. Ignace had a long history. For example, when Detroit, Mackinac & Marquette Railroad was selling over a million acres of land in the Upper Peninsula in 1882, it told potential purchasers that:

> Among the lands in the counties of Chippewa and Mackinac are large tracts of what are known as Burnt or Cleared Lands, the original growth of timber having been removed to a great extent by forest fires, leaving the land practically cleared, and requiring but little labor or expense to fit them for the plow.[7]

St. Ignace was not unique in thinking that farms would occupy once-forested land, whether cleared by fire or by the lumber industry. Across North America the expectation was that after the forest was cut down, agriculture would follow.[8] By the early years of the twentieth century, St. Ignace was actively recruiting farmers to settle in the area.[9] If townspeople thought about tourism, they conceived of it as a market for freshly grown farm crops. "This summer tourist trade is important [f]or the vegetables, fruit and dairy supplies [produced] for the various hotels" wrote the paper.[10] Linking farming and tourism had worked elsewhere.

For example, the settlers Muskoka, Ontario, which is located near Toronto and was an important tourist destination, received a steady income by selling their crops to cottage owners who came for the season.[11]

Farming was St. Ignace's future. Tourism, in the opinion of the local newspaper, was a useful way to make a few extra dollars. The tourists the paper first discussed were the hunters that came to the Upper Peninsula each fall. In late 1913, the paper noted that local businesses did little to encourage the many deer hunters who traveled into the Upper Peninsula to stop in St. Ignace. "This neglect is losing the town thousands of dollars annually," lamented the paper.[12] In 1916, a new city organization was formed to encourage the deer hunters, and the summer tourists coming to Mackinac Island, to spend time in St. Ignace. The organization's goal was to make more visible "our light hidden under a bushel."[13]

This was not the first time St. Ignace had tried to shine for tourists. In 1901, the Business Men's Association had published a thirty-eight page tourist booklet declaring St. Ignace "the Gate-way City to the Upper Peninsula."[14] In 1914, the Dunham House, which opened sometime before 1885 and was the city's first hotel, doubled in size and included a large "promenade," basically a covered porch, to be enjoyed during the warm season.[15] In 1915, St. Ignace promoters, likely thinking about how to attract people from Mackinac Island to their community, organized a Chautauqua. Chautauquas were a frequent summer event in many American communities. Originating in the Methodist summer camp movement, chautauquas brought together a series of programs that mixed adult education, motivational speakers, various forms of entertainment, and, often, a religious revival meeting into a single group of scheduled and ticketed events. The first of these programs was organized in 1874 alongside Lake Chautauqua, New York. From there, the idea, and the unique name, spread.[16]

The town's paper lobbied for holding a Chautauqua in St. Ignace.

> Little or nothing is done by the city to increase the attractiveness of the place or to furnish other amusements for the benefit of our summer visitors other than those furnished by nature. A little effort on the part of our citizens working together and in harmony could work wonders toward increasing our attractions and likewise the number of our summer inhabitants, and in

> the opinion of the Enterprise the opportunity for a beginning along these lines is furnished by the Chautauqua idea.[17]

That St. Ignace, despite the Business Men's Association publication of 1901, had done "little or nothing" to attract tourists was a regular theme in the *Enterprise*. The comments of an anonymous "summer sojourner," which the *Enterprise* published, laid out the problem. The sojourner shared that ten years earlier he had been about to leave Mackinac Island when, on a lark, he decided to visit St. Ignace. The scenery he saw while traveling on the ferry was enchanting. His enchantment, however, was short-lived. "Leaving the dock what a disappointment came to us. Alas, the beauty of the distant view was partially eclipsed by the rather dreary aspect of 'Main Street.' After a half hour's walk along this street, one remarked 'God made this a beauty spot—man defaced it.'" Despite this first impression, and for reasons the sojourner himself found inexplicable, he returned to St. Ignace. He came to "know and love" St. Ignace for its "reasonable rates, pure drinking water, healthful and appetite giving air, stores where one is well served and churches where a real Christian welcome is given." The sojourner found it a wonderful community but one that needed to "awaken to the possibilities of industrial expansion" through tourism.[18]

The *Enterprise* was convinced that the Chautauqua would expand tourism in the city. It did not. The city's first Chautauqua managed to turn a small profit, but a second Chautauqua held in 1916 was financially unsuccessful. Local merchants had agreed to pay for the event's expenses, should ticket sales fail to cover costs. When they were required to do that, enthusiasm for the event ended and there was no Chautauqua in 1917.[19] It also ended for several years any organized effort to attract tourists. The summer sojourner was always welcome, but Main Street remained bleak. Tourism was an afterthought.

What created a new interest in tourism was the collapse of the local economy that had been based on lumbering.[20] The community's population had declined by one-third between 1890 and 1920, from 2,704 residents to 1,852.[21] St. Ignace had always assumed farming would follow logging, but it had become increasingly obvious that farming was not viable in St. Ignace, or for the most part anywhere in the Upper Peninsula.[22] The soil was too poor and the growing season to short.

Facing this reality, St. Ignace's mayor called a town meeting in May of 1922 to spell out how he saw the future. He began with a hard truth. He told his fellow citizens that because of the "denuding of our forests," they "could no longer hope for any great advantage from the lumbering industry on which the town had for years depended." Leaving unmentioned the inability of agriculture to replace lumbering, the mayor told his audience that the community's only hope for future prosperity was to remake St. Ignace as "a summer resort that would attract the admiration of the country and thousands of visitors annually."[23]

To attract thousands of visitors each year, the mayor had a list of projects he wished to accomplish. Among many things, he wanted to build a short stretch of roadway to connect the downtown area to the main state highway, reconstruct the road leading to "famous" Sand Bay, convince the state to create a state-operated park in or near the city, clean up the city's waterfront and downtown, mark historical spots, improve the property around the community's mineral springs, create advertising material to promote the city, and open an information bureau to attract visitors. To do all these things, the meeting created a Boosters Association, with the slogan "Every citizen a booster." The newspaper jumped on the bandwagon, editorializing that every "Saint" should "BOOST."[24]

Key to the tourism initiative was that the state of Michigan would soon begin to operate an automobile ferry service across the Straits of Mackinac. Automobile owners had begun arriving in the straits area before World War I, but the trip almost always ended in Mackinaw City. The only way the first cars reaching the straits could cross over to the other side was to book a trip on the railroad ferry, but the process was neither convenient nor cheap.[25] The railroad ferry existed to moved railroad cars and "was not very eager for the [automobile] business."[26]

Getting visitors from Mackinaw City to St. Ignace was important because of the huge population difference between Michigan's two peninsulas. The 1920 US Census counted approximately 3.3 million people living in the Lower Peninsula, compared to approximately 333,000 individuals in the Upper Peninsula.[27] There was a consensus that making it easier

Indian Village, St. Ignace, ca. 1925. COURTESY MICHILIMACKINAC HISTORICAL SOCIETY, ST. IGNACE.

for the residents of the Lower Peninsula to cross the Straits of Mackinac would significantly boost all aspects of the Upper Peninsula economy. The state began operating an automobile ferry service across the straits, running between Mackinaw City and St. Ignace, in 1923.[28]

In 1924, the first full year of cross-straits ferry service, 38,468 vehicles made the crossing.[29] In 1930, the number of vehicles the state ferries carried had grown to 130,942 vehicles, with 221,136 people passing through St. Ignace, either northbound or southbound.[30] The new auto ferry service and the exhortations of St. Ignace's mayor led to a temporary interest in tourism. A state park with campsites was authorized in 1922, although it did not open until 1925.[31] In 1923, a privately owned information bureau came into existence, with the mission of promoting local attractions and businesses to visitors. A golf course was proposed in 1923. Although not mentioned by the mayor, some considered a golf course a necessity if St. Ignace was to compete against other resorts.[32] The golf course opened in 1926.[33] In 1923 the Gateway Improvement Company began to build summer cottages.[34] There was also talk that cottages would also soon rise on the "old furnace property" on the edge

Fort Algonquin, St. Ignace, ca. 1928. Courtesy Michilimackinac Historical Society, St. Ignace.

of town.[35] In 1927, "Evergreen Shores," perhaps the city's most elaborate rental cottage site, opened.[36]

History was also used to attract visitors to the city. Clarence C. Eby was a St. Ignace resident who saw the past as a way to make money in the present. Eby's most long-lived historical attraction was Indian Village, a tourist destination he opened with Duncan Bajbedauscen (aka Bobby Dosen), who was a Native American, which featured Native Americans making craft objects that were available for sale.[37]

Others also gravitated to this idea. For example, in 1928, Henry Norton opened his privately reconstructed fort, named Fort Algonquin, that was modeled on what he thought Fort de Baude looked like.[38] Another ambitious historical enterprise undertaken by the city to attract tourists in the 1920s revolved around the gravesite of Jesuit missionary Father Jacques Marquette. Marquette founded Catholic missions at Sault Ste. Marie and St. Ignace. He traveled widely and, while still quite young, died somewhere along Michigan's Lake Michigan shoreline. Native Americans who accompanied him buried him near where he died. They subsequently returned and carried his remains to the mission at St. Ignace. In the 1920s, the St. Ignace community began to hold an annual commemoration of Father Marquette's interment.[39] In 1927, the 250th anniversary of the return of Marquette's remains to St. Ignace, what had originally been a relatively modest, largely religious remembrance was reinvented as a grand public event.[40]

The new commemoration ceremony featured a pageant involving over four hundred local people. The John P. Rogers Producing Company of Fostoria, Ohio, which claimed to produce the largest outdoor spectacles in the United States, was hired to plan and operate the event. The pageant was a lengthy production that remembered not only Father Marquette but also included community history and straightforward entertainment. The script for the event included the arrival of a local beauty queen, a prophecy by Father Time, a ballet, an acknowledgment of the Native community, and reenactments of what the town's citizens viewed as important milestones, such as the arrival of the community's first European pioneer, the first wedding, and the first school. The grand finale brought the entire cast onstage, dressed as people from all over the world, who together sang the "Melting Pot Chorus."[41] In conjunction with the pageant, the largest parade held up to that time in the city's history was planned.[42] The dates of the Pageant were chosen to correspond with the three-day National Governors' Conference held on Mackinac Island as well as a meeting of the Michigan Historical Society, also held on the island. In addition, a community "homecoming" for those who had left St. Ignace but would like to visit their old friends, was also planned.[43]

This massive event, as well as earlier tourist-related programs, generated publicity about the community across the state. In 1925, the *Bay City Tribune* called St. Ignace a Cinderella city. Before the ferry, the town was simply a "jumping off point . . . that had to be endured because [it] could not be avoided." It had "unattractive stores and eating places." The *Bay City Tribune* noted that although "fair sleeping accommodations" could be found in St. Ignace, most people didn't look and "just got out of dodge." All that, the paper maintained, had changed. And best of all the community was not "squeezing too hard the goose who laid the golden egg." Tourists were not overcharged.[44]

Not everyone, however, had a positive view of St. Ignace. The *Cheboygan Democrat* wrote two years later, "St. Ignace has not yet awakened to the possibilities of this auto invasion and are making no effort to stop the tide there for a while in an effort to show them what a most wonderful and pretty place the Pointe is." The Cheboygan paper also described the St. Ignace merchants this way, "They are overcharging for almost anything that one of these travelers asks for."[45] When the editor of St. Ignace's newspaper described the claims of price gouging as "mostly pure bunk,"[46] the Cheboygan paper responded by printing examples of the high prices charged in St. Ignace. Taken aback, the St. Ignace newspaper editor wrote that, assuming what had been printed was true, such scoundrels would go out of business.[47]

But the *St. Ignace Enterprise* knew that the Cheboygan paper had captured at least a partial truth. St. Ignace faced serious challenges that were impeding its growth. In late 1923, the *St. Ignace Enterprise* printed a letter from a "friend" of the community who deplored "the lack of cooperation, the petty jealousy, the religious bickering and the growing formation of closed corporations and social cliques" within the community. Instead of "boosting," the community was "just drifting along."[48] A second letter to the paper, printed about two months later, said, "For a resort town that caters to the tourist trade during the summer months it [St. Ignace] is the most dilapidated place I have ever seen. . . . Don't the citizens believe in paint?" The letter also brought up the matter of prices, "you should treat the tourist right" and "don't overcharge him."[49]

Adding to concerns about St. Ignace's neglect of tourism was the often acknowledged problem of an unattractive waterfront, a problem made worse when local merchants constructed a "gasoline alley" with at least six gas stations near the State Ferry Dock.[50] The unattractive gas stations were well placed to capture business from cars coming off the State Ferry, but were also an invitation to gas up and "get out of dodge." The shortsightedness of the gas stations owners also found its way into city government. In 1929, Charles Eby, who had organized the information bureau boosting the city to tourists, closed it. The reason was that, after personally funding the service for many years, in 1928 he asked the city government to pay some of the costs. The city refused. As the newspaper put it, "The thing takes up more of his [Eby's] time than the Indian Village [Eby's successful tourist attraction] and 'he is through.'"[51] Eby was not the only booster finding that their hopes exceeded the community's support for tourism. A few years after it opened, the golf course was in financial trouble. In 1929, it was sold to an "angel" who, it was reported, would keep it operating.[52] The state park, while helpful in drawing visitors to the town, was also a frustration. The local community repeatedly complained the state was moving too slowly in developing the property.[53]

Tourism had not become central to the local economy. In 1932, the *St. Ignace Enterprise* published an editorial acknowledging the situation. Entitled "Let's Go! St. Ignace," the editorial prodded the community to embrace tourism. The arrival of the state auto ferry had given the city the opportunity to transform itself but, the paper concluded, no transformation had taken place. "It now remains for us to decide whether St. Ignace is to be just a gateway or whether it will be a point of real attraction in itself." Will it be "a place to buy gasoline or shall it be a DESTINATION?"[54]

In 1932, Mackinaw City, on the south shore of the straits, was also primarily a gateway, but it had begun to evolve into a destination. For much of the nineteenth century, the land that would become Mackinaw City was uninhabited. As a result of the War for Empire that had taken place during the eighteenth century, the British had taken over most of French Canada, including the small fort at the tip of Michigan's Lower Peninsula, Fort Michilimackinac. In the early 1780s, the British relocated the soldiers stationed there to a new fort that they constructed on Mackinac Island. Most people followed the soldiers to the island. Eventually, in 1857, a new community named the village of Mackinaw City was platted by Edgar Conkling and Asbury Searles, whose offices were in Cincinnati, Ohio. Conkling and Searles made extravagant claims about the future of the new community. "If one were to point out on the map of North America a site for a great central city in the lake region, it would be the immediate vicinity of the Straits of Mackinaw." Reflecting what they had likely read about nearby Mackinac Island, they added that Mackinaw City would become a "principal resort for pleasure and health."[55]

For over a decade after it was platted, the "great central city in the lake region" was uninhabited. In 1870, George Stimpson and his family became the town's first settlers. Stimpson built a dock from which he sold wood to fuel passing steamships. He would cut and haul approximately twelve to fifteen thousand cords of wood each winter and sell it from the dock during the shipping season, refueling around two hundred thirty steamers.[56] Eventually, other lumbering operations began nearby, and a small community developed around Stimpson's dock.

Although Conkling's claim that Mackinaw City would become a great city was hyperbole, Conkling was right in saying that geography would play a role in the city's destiny. Mackinaw City did not have a natural harbor and thus never substantially participated in lake commerce. However, it was the northernmost point of Michigan's Lower Peninsula and became the place where the Lower Peninsula's railroad network terminated.

The Michigan Central Railroad reached Mackinaw City in December 1881, coming north from Lansing through the center of the state, then swinging east to Cheboygan, where a booming lumber industry guaranteed regular freight traffic. From Cheboygan, the Michigan Central hugged the Lake Huron shoreline to Mackinaw City.[57] Mackinaw City residents celebrated the arrival of the railroad. But it was a small town, and a small celebration. The first steam engine coming into town was greeted by a bonfire. When those who had stood around the fire had sufficiently admired the train engine, they went to the town's sole tavern to continue the festivities.[58]

In 1882, the Grand Rapids and Indiana (GR&I) Railroad reached Mackinaw City from the west side of the state. The GR&I Railroad's main line largely following the Lower Peninsula's Lake Michigan shoreline.[59] In 1904, the Detroit and Mackinac (D&M) Railroad completed its route along Michigan's Lower Peninsula's Lake Huron shoreline to Cheboygan. There, its track connected to the existing Michigan Central Railroad track to Mackinaw City.[60] Train traffic was significant, and a ferry operated by the Michigan Transit Company, a business jointly owned by several railroads, was established to transfer railroad cars across the straits. The railroad ferry began operating in July 1882.[61] By the 1890s, three to five hundred railroad cars were being sent daily across the straits.[62] The ferry service ran year-round and carried both freight and passenger cars.

The railroads brought tourists to the straits, most of them bound for Mackinac Island. To accommodate these visitors, in the early 1880s, Mackinaw City developed a second ferry service that moved tourists from the railroad station jointly run by the Michigan Central Railroad and the Grand Rapids and Indiana Railroad to Mackinac Island.[63] The ferry boats coordinated their schedule with the arrival of passenger trains. In 1900, a person arriving at the Mackinaw City Railroad Station could be on Mackinac Island in as little as fifty minutes.[64]

As in St. Ignace, the goal of most people arriving at Mackinaw City was to leave. As one community resident recalled, "Mackinaw was like a getting off place."[65] Services for travelers that were made available were designed to meet the immediate needs of arriving or departing train passengers. A large dining room was built within the train station. On a busy summer morning, when many overnight trains arrived, the station restaurant could serve as many as eight hundred meals.[66] And summer was busy. In the 1920s, about twenty passenger trains arrived daily. The arriving trains were sometimes so long they extended beyond the city. One story was told of a person, traveling around 1900, who was riding in the rear of a train. The traveler asked the conductor why the train had stopped in the woods rather than in Mackinaw City. The conductor replied, "Lady, this is it."[67] While comical, the story had much truth in it. Otto Lang, who came to the straits area in 1922, recalled that when he arrived, Mackinaw City and St. Ignace were "mere stations" on the railroad line.[68]

If most people quickly left Mackinaw City, a few stayed. The industrious George Stimpson built what passed for the community's first hotel in 1874. It was nothing more than an addition built onto his house.[69] The 1883 Michigan Central Railroad's tourist guide, after first promoting the ease with which someone could leave Mackinaw City for either Mackinac Island or St. Ignace, noted that Wentworth House, which was attached to the train station and was owned by the railroad, was "one of the finest railroad hotels and eating houses in the United States." That claim was an exaggeration but did acknowledge the existence of another hotel in Mackinaw City.[70]

At the beginning of the twentieth century there was an attempt to create a summer community near Mackinaw City that would mirror the cottagers who went to Mackinac Island, although on a far more modest scale. Around 1900, Vine Harding inherited about three miles of undeveloped Lake Michigan beachfront property. He built a summer home and personally surveyed lots. Around 1907, he paved a walk along the beach and constructed a road approximately one thousand feet back from the beach walk. As described in a promotional brochure, Wawatam

Beach, "on the Straits of Mackinaw [and] commanding [the] finest view of Islands, beautiful scenery and passing ships," was "in the vicinity of the best fishing grounds in Michigan" and had "wonderful springs and health giving pine woods," all "within easy reach" of the railroads. The earliest recorded construction on a lot sold by Harding began about 1910.[71] In 1914, approximately forty cottages had been constructed.[72] The Wawatam Inn was constructed on the beachfront, offering a hotel for those who could not afford to buy a cottage.[73] Complimenting the Wawatam cottages was a second development, the Mackinac Straits Shores, which appeared on the Lake Huron side of the straits in the 1920s.[74]

As was the case in many Michigan lakefront towns, some cottagers came to love Mackinaw City. John Moore recalls arriving for the summer on the train from Kalamazoo:

> Those last six miles were the best and the worst, full of the long sitting with the bags ready and the coat over the plush back of the seat and my eyes watching, ticking off the landmarks . . . and at last the slow fall of the tracks cutting down the last glacial lake level to the flat where, with the suddenness of a dream, the train swung the wide curve into the first siding and there ahead was the wedge of lake and the old depot. . . . And in the first split moment, before the movement began again, our movements now, not the train's, all of the past summers came together and all the urgency to see it all at once, how the land had changed (and the fear that it had changed and wasn't the old dream anymore) and how the cottage looked and where the water level was this year—everything and everything.[75]

Although the railroads brought tourists to Mackinaw City they did little else to help the town's tourist trade. What a tourist saw stepping off the train in Mackinaw City was an industrial transit point. The railroads required large and unsightly freight yards near the center of the community to handle the many trains coming into and out of town, the majority of them freight trains. Adding to the clutter caused by a busy freight yard, the GR&I Railroad used track across from the train station, in plain sight of arriving visitors, as a dump for old and broken-down cars.[76] One glance from the railroad station's platform made it clear to someone stepping off the train that if they had come for picturesque scenery they had arrived in the wrong place.

The city's non-tourist-friendly economic activity was compounded by another problem tourists faced in Mackinaw City: there was not much to do. In 1913, a Detroit paper carried a letter from a New York resident, who commented of Mackinaw City, "This is an awfully dead place. The most exciting thing that occurs in the town is when the sheriff takes his prisoners down to the river for a bath."[77] In 1915, the city's government, somewhat grudgingly, agreed to name the town's streets to help tourists find their way around, although one member of the city council said that the town was so small this seemed "hardly worth while." "Nobody had any trouble finding himself in Mackinaw."[78]

Nor was the village government consistently concerned about the impression the community made on visitors. In 1919, Central Avenue, the city's main thoroughfare, was turned into a parkway. Sounding like a major improvement of the cityscape, the reality was far different from the image conjured by the word "parkway." The Central Avenue parkway had weeds in the middle and a sand rut on one side of the road.[79]

Tourism was an afterthought. As one resident recalled, during the 1920s, "most of the men worked for the railroads, fishing, or lumbering."[80] Fishing and lumbering were the less important industries in the town. Commercial fishing continued into the 1930s.[81] However the activity was considered largely unnewsworthy and thus there is little documentation of it. Lumbering was an important part of the local economy in the 1890s and early twentieth century.[82] In the 1890s, Mackinaw City was home to a shingle mill and a sawmill. Lumbering continued until after World War I, but, like in St. Ignace, it was in decline.[83]

Transportation was the heart of the town's economy. The local train station had thirteen employees. Outside of the station, the railroad employed thirty-seven workers in the "car department," thirty-one employees in the locomotive department, and a foreman and six workers ensuring the track in the station yard and nearby ferry docks was in good working order.[84] One longtime resident would reminisce that when his parents arrived in the community during the 1920s, of the seven to eight hundred people who lived in Mackinaw City, one hundred eighty worked for the railroad.[85] Even the town's children found a way to profit from the railroad. They made pocket money helping passengers carry bags onto departing trains. A kid could expect a fifteen- or twenty-cent tip for a few minutes work.[86] In addition to the workers and children in or around the train station, a large number of individuals worked on the railroad ferries, which, by 1915, were shuttling nearly sixty thousand railroad cars a year across the straits.[87]

In the 1920s, the state automobile ferry system, another part of the transportation network, became a major employer in the region. Each of the five ships running by the 1930s required a crew of about thirty-six, and the docks in Mackinaw City and St. Ignace employed another sixty workers on each side of the straits. Like the railroads, the state Highway Department created an unattractive dock site. The Mackinaw City dock had a parking lot accommodating 1,100 cars.[88] As originally built, the dock "was practically destitute of trees and shrubs."[89] Eventually, in 1927 or 1928, the state Highway Department did plant some trees and installed other landscaping around its Mackinaw City facility, but this was not a priority.[90]

Before the advent of automobile, tourism in Mackinaw City was primarily a small rural community where people made their living through transportation, fishing, lumbering, and in the summer by serving a tourist business largely created by transporting people to and from Mackinac Island. Like St. Ignace to the north, tourism was the least important element of the local economy.

CHAPTER 10

Automobile Tourism, 1912–1959

Automobiles created a boom in tourism across America and at the Straits of Mackinac. In 1925, the *Emmet County Graphic* summarized how it believed automobiles had changed Mackinaw City:

> With the development of the automobile and the evolution of the motor tourist taking place but a comparatively short time ago, the business men of Mackinaw City found it necessary to adopt themselves to changing conditions and today practically all commercial activity is aimed toward the tourist trade. . . . The business institutions of the town make every effort to serve and please the tourists who halt there on their way north or south.[1]

In 1929, the *Petoskey Evening News,* noted that drivers were doing more than briefly stopping on their way north or south. "Thousands who planned to 'go right on through' are so pleased with the village [Mackinaw City] that they stay here for extended visits."[2]

The railroads had brought the first tourists to the Straits of Mackinac's shoreline. As the automobile grew more common vacationers abandoned

Mackinaw City, Main Street, 1926. Courtesy Clarke Historical Library, Central Michigan University.

trains and turned to cars. In 1916, the author Theodore Dreiser explained why. "The railways have become huge, clumsy affairs, little suited to the temperamental needs and moods of the average human being." Automobiles offered "freedom from the shackles of railway timetables."[3]

If automobiles freed travelers from timetables, like the trains that had preceded them, they required an infrastructure, good roads instead of steel rails. In the first years of the twentieth century, intercity roads that cars could easily drive on were virtually nonexistent. To travel any distance, an automobile owner was required to navigate unconnected, rutted paths.[4] Roads going out of a town often led to nowhere because in nineteenth-century America, road construction was a county responsibility, and counties built their roads to bring farmers to the county's main town, or the nearest railroad station. There was little regard to how these county roads connected to one another at county lines. If someone planned to travel any distance, everyone assumed they took a train. But automobile owners, who had eschewed trains, demanded better constructed, interconnected roads. To meet that need, in 1913, the Michigan legislature enacted the State Trunkline Act, which established Michigan's first state highway system.[5] In 1921, Congress created a national highway network, with each federal highway identified by a unique number.[6] Tourism at the straits grew rapidly in the twentieth century because three US highways met in Mackinaw City. People from all over Michigan, and for that matter the nation, could drive to the Straits of Mackinac.

In the early years of automobile tourism, US 31 was particularly important to straits tourism. By 1925, US 31 made it possible to drive from the Indiana border along the Lake Michigan shoreline to Mackinaw City at a steady speed of thirty-five miles per hour, with bursts of up to sixty miles per hour in "open country."[7] More than a million people a year drove the road, two-thirds of them coming from outside of Michigan.[8] US 31 brought an unending stream of cars to Mackinaw City. Soon, the stream became broader as other federal highways came into the town, including US 23 running along the Lake Huron shoreline and US 27 that ran through the center of the state.[9]

Automobiles, however, not only brought more visitors, they also brought a different kind of visitor. Working-class people joined the upper- and middle-class tourists who had been coming by boat or train to the straits since the 1880s. As the twentieth century advanced, the nation's working-class families discovered they could afford to buy an inexpensive automobile. Most working-class families soon realized that a family car could be used to take a family vacation. This new group of people, however, were particularly sensitive to the cost of vacations. Hotels and restaurants were often not within their budget. That price sensitivity led them to bypass Mackinac Island but created new opportunities for both Mackinaw City and St. Ignace. Both communities embraced new, less expensive forms of vacationing: camping and overnight cabins.[10]

Mackinaw City's creation of a campsite occurred more by accident than design. The 1857 plat of Mackinaw City set aside about twenty acres of land thought to be the site of the old French fort to be a park. In the 1880s, the land was given to the village of Mackinaw City. Mackinaw City gave the land to the state of Michigan in 1903. The park land sat idle until 1910, when the state legislature assigned responsibility for it to the Mackinac Island Park Commission, mandating that 5 percent of the commission's money be spent in Mackinaw City but declining to add any money to the commission's state funding.[11]

In 1913, the state legislature finally offered the park commission some funding for its Mackinaw City park and the park commission began making small improvements to it. Among the improvements was a road for automobiles to the park and campsites. In 1920, a Grand Rapids newspaper reported that the park had "a big stone stove with iron top large enough for a number of parties of campers and picnickers to use at once." The paper added that picnic tables were located around the grounds and in a covered pavilion, supplemented by a "convenient pump" for water.[12] Over the next few years, playground equipment, a building in which to change into swimsuits, and electric lights appeared, making the park more attractive both to day visitors and overnight campers.[13] All this catered to the needs of the new motoring public, but none of it was unique. In 1925, Michigan was second in the nation in the number of state parks it had created, and by 1930, nearly nine million people visited them and campgrounds similar to those in Michigan were appearing across the nation.[14]

The most important thing the commission did to promote camping in Mackinaw City was something that they likely gave very little consideration to: hiring Chris Schneider to run the park and campsite. Schneider was appointed the park's caretaker in 1921 and continued in the job for more than thirty years. He had a personal style that endeared him to guests. He claimed to never forget a camper's face and he greeted campers who returned for a second stay like family coming home.[15] Schneider's pleasant personality was the epitome of the era's "mom and pop" campsites that were particularly successful at winning the hearts of campers.[16]

Park improvements and Schneider's gift for hospitality made the commission's campground in Mackinaw City popular. In 1924, the Mackinaw City campground issued 2,095 camping permits that brought 6,861 campers to the site. A total of 12,466 campers visited in 1927. Camping in Mackinaw City was so popular that, in 1923, the commission obtained title to twenty acres of land near St. Ignace in order to open a second campsite. The commission optimistically assumed that either the city of St. Ignace or the state government would pay to develop the vacant land. When neither did, the commission abandoned the project and turned the land over to another state agency.[17]

Fort Michilimackinac Campground, Mackinaw City, ca. 1920. Courtesy John Marsh Postcard Collection, Mackinaw Area Public Library.

The new highways brought ever more visitors with their tents, but the highway also brought many people who found camping uncomfortable. Many of these vacationers had a bit more money in their pocket, but not quite enough for a hotel. Yet they wanted their vacation to be something more than sleeping in a tent. Overnight cabins began to appear to serve these visitors. The cabins represented a transition between camping and a traditional hotel. Small and modestly priced, they were also called "cottages" or "tourist courts." Exactly where in America the first such cabin appeared will never be known with certainty, but some claim the first one appeared in Douglas, Arizona, in 1913 and the idea spread from there.[18]

First established in the straits area during the 1920s, most tourist cabins looked very similar to Sid's Red and White Cottages. Located in Mackinaw City, Sid's opened in 1930 with sixteen cottages, advertising that they were "all . . . screened to assure the guest that insects will not molest them during their stay." Sid's also had on the premises a gas station and a camp store and served chicken and fish dinners.[19] Each cabin

Sid's Red and White Cottages, ca. 1930, Mackinaw City. COURTESY JOHN MARSH POSTCARD COLLECTION, MACKINAW AREA PUBLIC LIBRARY.

operator spun a slightly different story to potential guests. For example, the Breakers Tourists Cabins, also located at Mackinaw City but adjacent to the state ferry dock, offered twenty cabins similar to those found at the Sid's Cottages but with "good fishing on the dock."[20]

Although these establishments tended to very similar, the published guides of the day assured drivers that "after a few days practice you should be able to tell pretty well from outside appearances whether the camp is operated in the manner you desire, and pass, with foot on the accelerator, those which are unkempt, dirty and down at the heel."[21] Visitors in Mackinaw City looking for a tourist cabin more often coasted past an unsatisfactory cabin to find one more to their liking. By 1937, Mackinaw City boasted over two hundred overnight tourist cabins and claimed more of them than any other town of its size in the world. St. Ignace was close behind in the number of cabins available to motorists.[22]

Tourist cabins represented one of the new forms of commerce that resulted from the nation's highway system, but many other enterprises were created to serve the motoring public, among them new restaurants built alongside the highways. One of the most remembered in Mackinaw City was the restaurant owned and operated by Harry Teysen. Teysen's family had operated a restaurant in Europe where Harry learned the business. When he immigrated from Luxembourg to the United States, first settling in Charlotte, North Carolina, he opened a restaurant. Around 1910, he came to Saginaw, Michigan, and eventually he became the food and beverage manager in the prestigious Saginaw Club. An avid sportsman and a victim of hay fever, Teysen and his family frequently vacationed in northern Michigan. His wife Grace had a hunch they could make a good living in Mackinaw City, and in 1926, they opened a small hamburger stand, located very near the State of Michigan ferry dock.[23] In about 1930, the Teysens converted the business into a sit-down restaurant with a dining room that could seat one hundred.[24]

Teysen knew how to please his customers and turn a profit. As one local put it, "The Teysen's have made Mackinaw City famous for hospitality, and have set a marked example of what can be done in developing the tourist industry which is Northern Michigan's pot of gold at the end of the rainbow." Or as another person put it, "He knew the service people enjoy and will pay for, and he knew the art of friend making."[25] Harry Teysen knew something else. One scholar summarized what Teysen seemed to know intuitively: "Roadside restaurants had to do more than serve food, they had to catch the attention of the passing motorist. Thus they often sought to portray an atmosphere of luxury, gaiety, or novelty." A good meal in an entertaining restaurant could fix a place in a tourist's mind.[26] To create a place that would be fixed in the tourist's mind, Teysen created "Teysen's Indian Curio Shop, Museum, and Dining," which in 1935 was advertised with the slogan "Travelers wise Teysenize."[27] Teysenizing occurred in a unique building that appeared to be a log cabin with four separate tepee roofs and a block house.[28]

Commercialization of the straits original residents was a well-worn marketing tool that originated in the nineteenth century. But like Clarence C. Eby's Indian Village in St. Ignace, Teysen's museum showed an unusual respect for Native culture. If Teysen's restaurant was unique in that it had a museum, it certainly was not unique among Mackinaw City businesses, many of which relied on clever names and roadside gimmicks to attract customers. For example, the Kewpee Doll and the Dockside Inn were not

Teysen's Restaurant, 1934, Mackinaw City. COURTESY JOHN MARSH POSTCARD COLLECTION, MACKINAW AREA PUBLIC LIBRARY.

far from Teysens.[29] And if Teysen's facade featured tepees and a blockhouse, Mackinaw City's Windmill Waffle Shop, according to a Petoskey newspaper, caught the eye of the traveler with "a genuine Dutch windmill and tower."[30]

Because Mackinaw City, and, to a lesser extent, St. Ignace, both enjoyed a large number of tourists driving into town, the communities' tourist industries fared well during the Great Depression. Automobile tourism was largely unaffected by the national economic crisis of the 1930s. In 1933, Americans spent almost as much money on oil, gas, and other automobile related expenses for vacations as they had in 1929, the last "good year" before the Depression. By 1935, automobile expenses for vacationing were at an all-time high. As one historian has noted, even when compared to other components of the recreation industry, including hotels, auto tourism fared "spectacularly well."[31] In Mackinaw City, new tourist cabins and boarding houses were being built. Business was so good that one source boasted that in the 1930s, Mackinaw City expanded and grew from a somnolent village to a burgeoning vacation spot.[32] Supporting this assertion were the number of tourists who visited the Mackinac Island State Park Commission's Mackinaw City site. In 1938, seventy-five thousand campers stayed at the campground, with another fifty thousand people stopping for a day visit.[33]

Mackinaw City also benefited from a major investment in the community by the Mackinac Island Park Commission. In 1929, the chair of the commission announced that the stockade that had once surrounded Fort Michilimackinac would be rebuilt. The project was not without problems, but on July 1, 1933, the reconstructed stockade was opened. The governor of Michigan presided over the event, and the public flocked to see the new attraction.[34] The stockade focuses attention on the fort's history, a story originally told by a small museum that Harry Teysen had assembled and installed in the campground concession stand he licensed from the commission.[35]

The stockade's construction led Mackinaw City's business community to experiment with historic tourism. On the same weekend that the fort's newly completed stockade was dedicated, the Mackinaw City Chamber of Commerce organized community residents to reenact the battle which occurred at the fort on June 2, 1763. The battle was part of a larger Native rebellion against the British, named Pontiac's Revolt.[36] In the words of the event's organizers, it was a "celebration of the 170th anniversary of the massacre of the entire British garrison of Fort Michilimackinac at Mackinaw City."[37] As was traditional with any Michigan summer festival, the organizers attempted to attract a larger audience by including a wide range of entertainment. Visitors with little or no interest in history could enjoy a parade, two band concerts, boat races, boxing matches, and an evening street dance.[38] The weekend of the stockade's opening and the pageant attracted fourteen thousand individuals to Mackinaw City.[39] The pageant was successful enough in attracting visitors to be repeated the next year.[40]

The reconstruction of Fort Michilimackinac's stockade in Mackinaw City led to hopes in St. Ignace that the French fort that had once stood on the north shore of the straits could also be remembered and rebuilt. However, the Mackinac Island Park Commission, which paid for the fort stockade construction in Mackinaw City, had no interest in rebuilding Fort de Baude. The most frequently heard explanation for this was that the commission did not own land in or near St. Ignace, thus making the project impossible. But even when land was offered at no cost, the commission was not interested. At their meeting of September 27, 1935, the commission declined to become involved in a proposed park project in St. Ignace for which land was available. The commission determined the project was "just more expense and work."[41]

If St. Ignace was to have a tourist attraction focused around the fort that had once stood in or near their city, the residents of the community would have to do it themselves.[42] As a first step, in September 1930, the local historical society promised to place a monument at the site of the fort.[43] By March 1931, the project had expanded to encompass purchasing the land where the marker was to be put in anticipation of reconstructing the fort.[44] The idea of a land purchase and reconstruction of the fort was problematic because no one truly knew where the old fort had stood.[45] There were educated guesses, but they remained guesses. Curiously, this obvious concern did not slow discussion of the project.

In August 1933, a public meeting was held in St. Ignace to promote "civic betterment" a phrase that largely meant attracting more tourists to the community. The meeting resulted in the organization of a Chamber of Commerce, which, in 1934, took the lead in rebuilding the fort, most likely in the city's Marquette Park.[46] But as 1934 moved forward, the chamber's plans began to fail. The St. Ignace newspaper described the "heroic attempt" of the city's Chamber of Commerce to "thwart the bickering sectionalism which to date has blocked many efforts toward improvement of the condition of St. Ignace as a tourist town." The paper put the best spin possible on the situation, concluding that the chamber would carry the day.[47] The chamber did not carry the day.

The collapse of the chamber's effort left the fort project and efforts to spur tourism in general dormant but not dead. In 1936, a revival of the fort project began when local historian Father John Howard reported he had conducted exhaustive research and had the necessary information to rebuild the structure.[48] In 1937, the newly organized St. Ignace Lions Club took on the project of rebuilding the fort.[49] As had happened in 1934, the project faltered.[50] Under the headline "Committee Flays Civic Indifference to Fort Project," the St. Ignace newspaper shared the Lions Club's concern over the "lack of enthusiasm" and "apparent indifference" shown by the public. The Lion's Club leadership was quoted as saying, "We must center our efforts upon the historic value of this vicinity to entertain people," to which the paper added words similar to those it had written in 1934: "We must all work together and contribute our share in planning for the future development of this community as a resort center."[51] As had happened in 1934, the fort project failed, and, more broadly, St. Ignace did not become a resort center.

Although the failure of the proposed reconstruction of Fort de Baude marked the most public example of St. Ignace's failure to become a tourist center, there were other examples of the "bickering sectionalism" that slowed the development of tourism in the community. In 1930, the state proposed to relocate its St. Ignace ferry dock a bit more than a mile from its original location in downtown St. Ignace. The governor knew there would be local opposition to the plan and tried to answer opponents' questions before they were asked. First, he personally promised that the state highway would continue to funnel traffic from the new dock through downtown St. Ignace, reassuring the local business community that the traffic the ferry had brought them would continue.[52] Second, the governor promised that unlike the current ugly dock, the new one would charm visitors with the "beauty of the landing place and its surroundings." This, he suggested, would encourage ferry riders to stay in St. Ignace.[53] The governor's words, however, had little impact. A dozen or so business owners immediately protested that relocating the dock would ruin St. Ignace. The town's mayor joined these business owners in opposing the dock relocation, as did one of the two newspapers then publishing in the community.[54]

The self-interest of the opposition in St. Ignace to the new dock was called out by others. The Cheboygan newspaper reminded readers that the old dock site was ugly, presenting "everything but an inviting appearance."[55] The paper in Sault Ste. Marie condemned the St. Ignace business community, writing that "the state ferry is not the property of St. Ignace and its oil station men." The Sault Ste. Marie paper may have gone out of its way to touch a local sore spot, when it added that if St. Ignace really wanted to attract tourists, it needed to clean up its waterfront and offer visitors "something besides gas and oil."[56]

The dock controversy of 1930 showed that part of the community that would rather make money selling a tank of gas rather than work

Downtown St. Ignace, ca. 1925. Courtesy Mackinac Island State Park Commission.

toward "the future development of this community as a resort center." A subsequent controversy showed the passing of ten years had done little to change things regarding developing what the Sault Ste. Marie paper had called an unattractive waterfront. In 1940, the city purchased three hundred feet of waterfront property. It announced the land would be used to create a park. A committee was appointed to design the park and, more broadly, create an "orderly plan" for city beautification. The state highway department became involved and developed a plan that included a fountain featuring colored lights for evening illumination.[57] It seemed the city would finally begin the often-discussed transformation of its waterfront into a tourist attraction. But a careful reader of the local news would know that there was an additional agenda afoot. In addition to a new waterfront park, the city was hoping some of the waterfront property would house a factory to process and freeze fish to be sold in grocery stores.[58]

About a year after the initial land purchase, the mayor announced the city's intention to sell 180 of the 300 feet of water frontage to a company that would build the frozen fish plant. The mayor focused on the economic advantages the enterprise would bring to the city. "There will be a big gain to our tax roll and in labor employed." Because of this, the mayor said, "We should offer the corporation every inducement we can to enable them to operate as soon as possible." There was no dissenting voices among the city's other elected officials. The city council voted unanimously to sell the land.[59]

The mayor had the good sense not to mention the factory's impact on tourism, but other public officials was less restrained. The city attorney said, "The plans, if in operation, would bring inestimable advertising to the Straits country." This was because "the entire business is the type to excite the imagination of the public who would remember the St. Ignace products and be interested in the locality."[60] The idea that the few consumers who read the label on a box of frozen fish and learned it had come from St. Ignace would then visit St. Ignace seems farfetched. More likely, the existence of a smelly factory cleaning and freezing fish on the city's waterfront would be another reason to fill up the car's gas tank and drive out of town. The controversy over the fish plant was short lived. The factory opened in 1941 and the company that owned it was bankrupt the next year.[61]

Mackinaw City could be equally divided over how to expand the local economy, and equally willing to sacrifice the area's natural beauty in return for a profit. In Mackinaw City, the question was the best way to expand automobile tourism and a repeated battle over constructing a new shoreline road that would connect Harbor Springs and Mackinaw City. The proposed eighteen-mile shoreline highway would run through Wilderness State Park. Proponents of the highway claimed it would bring an extra fifty thousand automobiles to the area each year but left unstated the amount of new business that they hoped would result from the new traffic. But those opposed to the road reminded everyone that the park represented "the last uncommercialized area below the Straits of Mackinac" along the shoreline of the Great Lakes. Building a new state highway through it might draw fifty thousand new automobile drivers to the straits, but Wilderness State Park was already drawing forty-five thousand visitors annually. It seemed that the road would create, at best, only a marginal increase in tourism. One could argue that it actually would hurt tourism, if the estimate offered by the road's supporters traffic estimate was questioned, and one considered what might happen to park attendance if the park's shoreline scenery was dramatically changed by a new road. The proposal was made three times between 1928 and 1938 and was rejected each time by the State Conservation Committee, which controlled access to the land.[62]

Discussions about tourism that frequently occurred in both Mackinaw City and St. Ignace during the 1930s came to an end with World War II. Gasoline rationing during the war caused a steep decline in tourism in both communities.[63] When the war ended, both Mackinaw City and St. Ignace were well positioned to expand their tourist trade. That growth was particularly facilitated by a fundamental change in working-class life. Between 1930 and 1940, the majority of America's industrial workers had been granted a paid vacation. The idea of paid vacations for working-class employees had been discussed since the beginning of the twentieth century, but it was only in the late 1930s that most factory works had received a paid vacation. The travel restrictions of World War II had given factory workers little opportunity to use their newly granted employee benefit for a trip, but at war's end, they were ready to get in the car and go.[64] These working-class travelers joined with the middle class and upper class, all of whom had been deprived of vacations during the war years, to make tourism boom everywhere in America.[65] As one author put it, "With the three things that had become symbols of the good life after the war—a job, a car, and a vacation—Americans hit the road in record numbers."[66] So many people came to Mackinaw City that in 1947, the town projected a 40 percent increase in tourism. Mackinaw City held its first big postwar celebration in 1947 during the July 4 holiday.[67]

Prospects were so bright that Mackinaw City experienced a building boom. The Mackinaw City village clerk announced in May 1947 that fifty-six new building permits had been issued, in a town with only 422 existing buildings. It represented a 20 percent increase in the assessed tax value of the community. But the sound of saws and hammers could not change old realities. Mackinaw City remained both a transportation center for upper- and middle-class travelers on their way to Mackinac Island and a tourist destination for working-class families. Symbolic of the old ways, the town's single largest building project was a new Greyhound Bus station located opposite the state ferry dock.[68] But like the merchants on Mackinac Island who tended to focus on the happy sound of money entering the cash register, Mackinaw City's tourist industry in 1948 seemed delighted when it was estimated that two thousand cars per day stopped in the city, with charter bus service up 17 percent over the previous year.[69]

St. Ignace also saw the possibilities of an increased tourist trade. In 1947, the city organized a Junior Chamber of Commerce, which picked up promoting tourism where the Chamber of Commerce and the Lions Club had left off. "Many reforms are necessary," said the Junior Chamber of Commerce. Among them, "the cooperation of every individual must be sought in the creation of good will towards St. Ignace. As the gateway to the Upper Peninsula, we are directly responsible for that all-important first-impression the tourist carries with him through the whole of northern Michigan. Our dealings with these pleasure-seeking people should be in SERVICE not in EXPLOITATION."[70] As a first project, the Junior Chamber of Commerce organized a city clean-up campaign, with the enthusiastic support of the mayor.[71]

What appears to be St. Ignace's first major postwar event was held over the weekend of August 13–15, 1948, to celebrate the twenty-fifth anniversary of the state auto ferry service. The event included a parade, "anniversary exercises," a moonlight boat excursion, and a "Queen's Coronation Ball, celebrating the selection of a Miss Straits of Mackinac." Also included was a swim from Mackinac Island's British Landing to the ferry dock at St. Ignace.[72] If it appeared that St. Ignace had finally come to value tourism, the city's long troubled waterfront suggested otherwise. In 1947, St. Ignace welcomed a furniture manufacturing plant to the community waterfront.[73] The struggle continued between unsightly industries on the waterfront and a redeveloped waterfront focused on tourism. Despite this, an observer would likely conclude that whatever the people on either side of the straits were doing about tourism, it was making them money. This optimism was encouraged by a new governor of Michigan elected in 1948, G. Mennen Williams.

Governor Williams fulfilled a long-existing pledge by the state's Democrats to build a bridge connecting Michigan's Upper and Lower Peninsulas. The conventional wisdom was that the opening of the Mackinac Bridge would lead to a new era of tourism at the straits and throughout northern Michigan. The *Detroit Free Press* wrote in 1954, when the bridge was funded and construction would soon start, that "the bridge promised rich reward to the state. It will open a news business and tourist doorway to the northwestern section of the Nation."[74] A few years later, Governor Williams would add his "Amen" to this hope, promising the bridge would annually bring an additional $100 million to the region's tourist industry.[75] At the straits, the mood reflected the optimism expressed in the Detroit newspapers and by the governor. A 1953 study "paid for by businessmen on both sides of the Straits" predicted the bridge would create a huge economic boom.[76]

Optimism, however, was not equally distributed between St. Ignace and Mackinaw City. In St. Ignace, the mood bordered on euphoria. When the bridge opened in 1957, it was claimed property in the eastern counties of the Upper Peninsula had doubled in value, while property values in St. Ignace had increased tenfold.[77] The boom in property value was seen as the harbinger of more good things. The secretary-manager of the St. Ignace Chamber of Commerce predicted that "in another five years, St. Ignace will have doubled in size," claiming that this estimate was not just his opinion but based on "a survey we had made of what to expect."[78] To get ready, the St. Ignace Chamber of Commerce, in 1956, created a five-point plan for "community betterment," one of which was the often-discussed hope to beautify the city's waterfront.[79]

If there was a concern in St. Ignace, it was that the bridge would bring too much of a good thing. Prentiss Brown, chair of the Mackinac Bridge Authority and a longtime St. Ignace resident, worried that the flood of tourists would transform the area in undesirable ways. We "do not want it [St. Ignace and the Upper Peninsula] to become a Coney Island littered with empty bottles, cans, and popcorn boxes, nor do we want private parties to control vast acreage from which the public would be barred."[80]

Despite the unbridled enthusiasm within the community, a voice of concern was occasionally heard. For example, in February of 1958, a representative of the Michigan State University extension program spoke to an audience of approximately two hundred about tourism. Although upbeat about the number of people soon to visit the straits, the speaker noted that newly planned freeways, which would follow the bridge's construction, would allow people to go "whizzing" past, a problem the St. Ignace's business community had to develop plans to "combat."[81]

The expectations in Mackinaw City were more mixed. Many of the town's citizens also assumed that the Mackinac Bridge would be good for tourism. To prepare themselves, in 1955, the Mackinaw City merchants reorganized their town's Chamber of Commerce.[82] But some Mackinaw City business owners, reflecting the thoughts expressed in St. Ignace by the Michigan State University extension speaker, worried that while the new bridge might bring more tourists to the area, fewer of them would stop in Mackinaw City. Furthermore, they reminded the optimists that despite the efforts to make Mackinaw City a tourist destination, much of community's tourist economy remained based on it being a transfer point between Michigan's two peninsulas. What the contrarians focused on were the business opportunities that would be lost when tourists no longer had to kill time waiting for a sailing of the state auto ferries.

That wait created lucrative business opportunities. Vendors roamed the state ferry dock area. "Motorists are accustomed to mass rushes by ice cream, smoked fish and newspaper peddlers."[83] Retail property near the dock was highly valued, since people waiting for a ferry could conveniently walk across the parking lot for a sweet treat, a bite to eat, or perhaps a souvenir. Subtly acknowledging this business model, after World War II, Teysen's Restaurant, which was located near the dock, added cafeteria service, making it possible to more quickly serve people. This was an important consideration when potential guests were deciding whether they had time to eat before the ferry boat would leave.[84] Even local charities took advantage of the wait for a ferry. In 1953, Mackinaw City's First Presbyterian Church's Women's Group baked and sold approximately 1,200 doughnuts to deer hunters waiting for a ferry to take them to the Upper Peninsula.[85] A bridge might bring more travelers, but how many would stay on the road and drive directly to the Upper Peninsula, eliminating, among other things, the church ladies' fundraising opportunity?

If the ultimate impact of the Mackinac Bridge on local tourism could only be guessed at, everyone in the region agreed that a huge celebration to commemorate the bridge's construction, actually several celebrations, would be good. The parties began with the groundbreaking ceremonies. On May 7, 1954, groundbreaking for the northern terminus of the bridge was held in St. Ignace. An early morning snowstorm almost led to the event's cancellation and greatly lessened the crowd, but the ground was broken. The next day, the sun shone in Mackinaw City where twenty-three bands, seventy-five floats, fireworks and a street dance celebrated groundbreaking for the southern terminus of the bridge.[86]

The Mackinac Bridge opened for traffic on November 1, 1957. Although the weather at the straits had been terrible in the last few days of October, "a brilliant sun bathed the Straits of Mackinac for this historic opening of the bridge."[87] A total of 150 reporters covered the day's events.[88] The official ceremony began at 11:00 a.m., when Governor Williams led a convoy of cars a mile long that drove over the bridge from St. Ignace to Mackinaw City. After lunch, the governor led the official convoy back across the bridge to St. Ignace.[89]

In addition to the ceremonial first drive over the structure, the day saw the final voyage of the state auto ferries. As part of the terms under which the bonds that financed the bridge were sold, the state had pledged to end the ferry service when the bridge opened. As November 1 approached, newspapers reported that people were taking "sentimental journeys" on the state ferries.[90] On the afternoon of November 1, the fleet's newest vessel, the *Vacationland*, made the services' final crossing.[91] The dignitaries who had driven in the motorcade earlier in the day also attended this "near tearful" event. Those aboard the ship were issued souvenir tickets. Many who did not have a ticket came to the dock simply to see the ship's last scheduled departure. In Mackinaw City, the *Vacationland* sailed away to a "thunderous salute" given by well-wishers in parked cars who sounded their horns.[92]

Although the governor and his mile long convoy opened the bridge on November 1, the idea of dedicating the bridge as winter set in struck everyone as a bad idea. Instead, a three-day event, described as "the greatest celebration ever held in Michigan,"[93] was planned for the last weekend in June 1958. Advanced events were planned on Thursday, June 26, in St. Ignace and on Friday, June 27, in Mackinaw City, with the formal dedication to take place Saturday, June 28. Plans were also made for additional events in nearby Cheboygan.

The greatest celebration ever held in Michigan required the attendance of beauty queens from each of Michigan's eighty-three counties. If the queens brought their beauty to the straits, the military brought its brawn. A total of 2,500 military personnel were sent to the area. The equipment they brought with them included a Navy destroyer escort, five Navy patrol craft, three Coast Guard cutters, and several Coast Guard helicopters that were to demonstrate air-sea rescues. Not to be outdone by events on the water, the Air Force planned flyovers by military aircraft.[94] Even the United Sates Post Office was involved. It issued a commemorative stamp celebrating the bridge's dedication, which would go on sale June 25. The postmaster of Mackinaw City said in May he had received two hundred thousand orders for first day covers.[95]

If outsiders had a large part in the dedication, the local communities also planned big events. The biggest were huge parades in both St. Ignace and Mackinaw City. The St. Ignace parade featured fourteen bands. Not to be outdone, the parade planners in Mackinaw City recruited twenty marching bands and 140 "units." Perhaps the most unique unit in the Mackinaw City parade was a float carrying a young Mackinaw City girl with a special connection to the day's celebration. Born on the day ground was broken to begin bridge construction, May 7, 1954, Bridget had been the talk of Mackinaw City. In 1958, four-year-old Bridget rode on her own float in the town's parade. There were so many activities that some of them had to be located in Cheboygan, approximately sixteen miles away from Mackinaw City. Planners erected a forty acre tent city in that community where a 4-H show was to be held. Cheboygan also planned yet another parade and a banquet.[96]

It was to be quite a show, until the weather changed everything. Seventy mile and hour winds lashed the straits the morning of June 26, grounding aircraft, wrecking exhibits, blowing over tents, and marooning G. Mennen Williams in the Governor's Residence on Mackinac Island.[97] St. Ignace, which had to deal with a fierce snowstorm when the bridge groundbreaking ceremonies were conducted, held the structure's dedication party in that city "over streets littered with tree branches and debris from the big wind."[98] But the weather eventually cleared and the ceremonies and celebrations were conducted. Arthur Juntenen, a *Detroit Free Press* travel writer, summed up the weekend this way.

> They tied the ribbons, glorified the bathing beauties, paraded everything that moved, talked themselves hoarse, over-ate and over-drank and left the deck hands to sweep up the confetti. . . . The seagulls are back temporarily having a ball on leftover hot dogs, hamburgers, popcorn and peanuts.[99]

The seagulls feasted on the trash of the estimated 100,000 to 250,000 people who had attended all or part of the three-day event.[100] As for the longer term, the *Detroit Free Press* summed up its opinion about the future in an editorial cartoon showing cars laden with camping gear driving across the bridge under the headline "Tourist Business."[101] Time would tell if the expectations expressed in the newspaper's cartoon, and by so many others, would be met.

CHAPTER 11

The More Things Change, 1960–1990

Optimism was the watchword when it came to tourism in St. Ignace and Mackinaw City in the first years after the Mackinac Bridge opened. In 1960 *Detroit Free Press* columnist Judd Arnett expressed that mood.

> [The Mackinac Bridge's] effect upon the economies of both the Lower and Upper Peninsulas is obvious. I had remembered Mackinaw City and St. Ignace as sleepy, dormant communities gradually using up the remaining lengths of shoestring pasts, but today they burgeon with new motels and assorted other conveniences for the tourist, and the end of this upsurge is not in sight.[1]

In 1963, another Detroit *Free Press* writer claimed one thousand new hotel rooms had been added in Mackinaw City since the bridge had opened. Although it is likely the one thousand rooms mentioned represented the total number of rooms in the city rather than new hotel rooms, the reporter went on to describe Mackinaw City this way:

> There is fresh paint in evidence, ferries shuttle gayly out to Mackinac Island and the motel, restaurant and souvenir shop proprietors have that beaming, light-hearted attitude that comes with cash in the till.[2]

In the evening the city's "motel row" had a "glitter equaled only by Las Vegas."[3]

Reporters were not the only ones who believed a regional economic expansion was occurring. Many local residents were also optimistic. St. Ignace and Mackinaw City had more people living in them than ever before. In 1960, 3,334 people resided in St. Ignace and 1,023 people lived in Mackinaw City.[4] The two community's growth encouraged new investment. For example, George Stemkowski was selling insurance in Mackinaw City, but he felt that the Mackinac Bridge made possible a new business opportunity. In 1962, he rented property near Fort Michilimackinac and opened a hot dog stand, where he sold chili dogs made from a recipe given to him by his father-in-law, who owned a bar in Detroit. A year later, he bought the property and built and opened the Fort Restaurant & Gift Shop.[5] Others were equally enthusiastic. As one local hotel owner, referring to the then under-construction I-75, said, "How could a swanky new-super-duper highway do anything but good?"[6] Another merchant called the road a "stream of gold."[7] Although it was never expressed in so many words, it seemed that Mackinaw City and St. Ignace might come to be widely seen as tourist destinations in their own right, rather than mere gateways to Mackinac Island.

By 1964, however, the economic optimism that was born from the construction of the Mackinac Bridge and the interstate freeways had begun to wane. Business was not what had been hoped. Newspapers reported that the tourist economy of the entire Upper Peninsula, after initially growing about 10 percent when the bridge opened, had first stabilized at that higher level, then declined by 26 percent.[8] But the straits tourist industry had always had its ups and downs. That things did not look bright in 1964 did not deter those who had spent decades at the straits. Among those who kept the faith that better days were to come was W. Stewart Woodfill and the staff of the Mackinac Island State Park Commission.

One of the projects Woodfill endorsed was rebuilding Fort de Baude on the straits' north shore.[9] Woodfill, however, conceived of the project differently from how it had been thought about in the 1930s. Instead of rebuilding the fort on its original site based on archaeological or historical research, Woodfill wanted to rebuild Fort de Baude where it could be easily constructed and where it would be easily accessible to visitors. Publicly owned land adjacent to the heavily trafficked Mackinac Bridge toll booths was the site he selected. Woodfill, however, was vague about the project details and people began to ask questions. Eventually, Woodfill took the unusual step of agreeing to speak at a meeting in St. Ignace, where he likely hoped he could sweep away the public's concerns.[10]

The meeting began with a presentation by Eugene Petersen, assigned the job of dealing with the issue raised by some that the site Woodfill proposed was demonstrably not where the fort had originally stood. Petersen pointed out that no one knew exactly where the Fort de Baude had stood. Time and money would be needed to find the site and map out the fort's boundaries. The archaeological dig would be challenging because the best guess was that the fort had stood in what had become downtown St. Ignace. If that was where the fort had been, then complicated land issues might make rebuilding the fort there impossible. Petersen's presentation ended in polite applause.[11]

Woodfill delivered a presentation focused on money. He discussed the revenue the commission received from restored Fort Mackinac. Said Woodfill, "It is an oil well steadily pumping every day to generate money for history and park projects." He then explained how a similar oil well would result from a reconstructed Fort de Baude near the Mackinac Bridge toll plaza. Woodfill made several points to justify constructing the fort at this location. No archaeological investigation was necessary because there was no reason to believe the original fort had stood at that site. The site was already public land, thus simplifying land issues. Without the need for archaeological work and with easy access to the land, work could begin quickly. Finally, when the construction was done, the fort would be easily accessible to traffic, and although Woodfill may not have mentioned it, with plenty of free parking. "It was a magnificent selling job," remembered Petersen.[12]

His sales pitch over, Woodfill said Petersen would answer questions. A well-prepared man stood to speak, who undid the spell cast by Woodfill. The rebuttal came in the form of several questions. Th first question alluded to what Petersen had already said about the original site of the fort but went in a different direction. Before spending money near the Mackinac Bridge, why not find the original site and explore the possibility of reconstructing the fort there? Petersen might be right in everything he had said. But Petersen was also just guessing. Why not make a decision from verified facts?[13]

Next, the man said that he was aware of concern about the entrance fee now charged at Fort Mackinac. Wherever the reconstructed Fort de Baude was located, it would be better if admission was free.[14] Woodfill knew this was impossible. Woodfill's motive for building at the north end of the Mackinac Bridge was because it represented the location that would likely generate the most money, money needed to pay back the bonds that would fund the project. Oil wells needed to be placed in the right spot.

Finally, the questioner said that while he could not speak for St. Ignace's municipal government, he was certain the city council would want to be involved in the project's details. He was likely right, and Woodfill hated committees and their inevitable, time-consuming debate. He was not about to deal with a city council questioning his every move. Petersen called it a "sobering rebuttal." Woodfill knew he had not carried the day. With a "painted smile" on his face, he and Petersen left the meeting. Woodfill's presentation in St. Ignace ended his interest in rebuilding Fort de Baude.[15] Although the commission would continue to explore the possibility of a fort or museum project in St. Ignace for several years, nothing would come of it.[16]

What led to a serious effort by the commission to build a museum on the straits' north shore was a 1983 investigation by a state legislative committee into the relationship between the Mackinac Island State Park Commission and the area's local governments. The hearings held by the committee made clear that festering wounds existed between the commission and St. Ignace's municipal government. Testifying before the committee, St. Ignace's city manager was diplomatic, but direct in his comments. He said that there was a sense in his community that the commission had never had much interest in the straits' north shore and that the situation was getting worse. "Many people in St. Ignace, however, feel the Park Commission is moving away from us." To put to rest the commission's frequently voiced claim that it lacked the land and the money to carry out projects in St. Ignace, the city manager pointed to the park commission's Mill Creek project. Several miles east of Mackinaw City, the Mill Creek project required both the acquisition of a large amount of land and the investment of a good deal of money.[17] The city manager added that to compensate for the lack of involvement by the commission, St. Ignace had begun working on historical tourism itself through the community's Downtown Development Authority (DDA), which had taken on projects at the Marquette Mission site and the site of the Huron Indian Village.[18]

Petersen, testifying on behalf of the commission, did not question the city manager's assessment that the commission had little involvement on the north shore of the straits. However, after disparaging the work of the DDA, he laid the blame for the lack of involvement on the leadership of St. Ignace. Petersen said that the commission might be willing to invest in the community, if there were a significant piece of land with which to work, not "a few small city lots in St. Ignace."[19] Petersen also reminded the committee that a few years earlier, St. Ignace's city government had failed to support the transfer of a nearby state park to the commission. St. Ignace, Petersen concluded, "was not always agreeable to MISPC [Mackinac Island State Park Commission] overtures to develop new areas."[20]

Petersen's rebuttal persuaded some. One member of the legislative committee agreed that the problem lay with St. Ignace, not the commission. However, Petersen's effort to place the blame for the bad blood squarely on St. Ignace's leaders was undercut when a member of the park commission who was in attendance was heard to say "I wanted to get a St. Ignace plan going but there were two factions and mine didn't win."[21]

Although the legislative investigation did not lead to immediate action, its recommendation that the commission begin work in St. Ignace "at the soonest opportunity" did get people thinking.[22] What this led to was a proposal to construct a transportation history museum. As early as 1964, Eugene Petersen had mentioned the possibility of transportation museum at the north end of the bridge. St. Ignace officials also expressed interest, but no work on the project was begun at that time.[23] But in the 1984, the Mackinac Island State Park Commission staff took the lead in planning for such a project,[24] funded by the Mackinac Bridge Authority. A study released in 1988 envisioned an almost eighty-three-thousand-square-foot facility that would include a transportation museum, gift shop, restaurant, and highway welcome center. The building would sit on a twenty-seven acre site and cost an estimated twenty-one million dollars to construct. The study estimated 150,000 paying visitors a year would stop at the new museum.[25] In 1990, a state appropriation made possible the creation of preliminary schematic plans for the building.[26]

The building was never built because the best estimates concluded that annual operating costs would exceed annual income. The project ended when it became clear that it would not be financially self-supporting. In the end, only a highway welcome center was built, paid for and operated by the Michigan Department of Transportation, which funded several similar welcome centers through its annual budget.[27]

Although the commission never authorized a large project in St. Ignace, the commission spent generously on developing projects in and around Mackinaw City. It is fair to say that no agency, public or private, did more than the commission to make Mackinaw City a true tourist destination. That work began with the reconstruction of Fort Michilimackinac. On Woodfill and Petersen's return to Mackinac Island from the St. Ignace meeting regarding Fort de Baude, a trip that involved driving from St. Ignace to Mackinaw City and then taking the ferry from there to the island, Petersen talked up a project in Mackinaw City that he wanted.

> It seems they [the St. Ignace audience] were not enthusiastic for your program. Why don't you consider doing something in Mackinaw City. If archaeologists can find Fort Michilimackinac's foundations, I believe a credible reconstruction could be done. This can be exciting, and besides you already control the site. He [Woodfill] looked at me [Petersen] steadily and said, "I doubt if we can sell bonds to do anything in Mackinaw City. But we're over here and might as well take a look."[28]

Taking a look undoubtedly complicated the task of persuading Woodfill to undertake the project. By the 1950s, the stockade that had been dedicated in 1933 had deteriorated so badly that some park staff had suggested it be demolished and the area turned into a parking lot.[29] Despite this, the restoration of Fort Michilimackinac was begun. Woodfill likely hoped against hope that, as Fort Mackinac on Mackinac Island was, Fort Michilimackinac

Fort Michilimackinac, ca. 1960, Mackinaw City.
COURTESY CLARKE HISTORICAL LIBRARY, CENTRAL MICHIGAN UNIVERSITY.

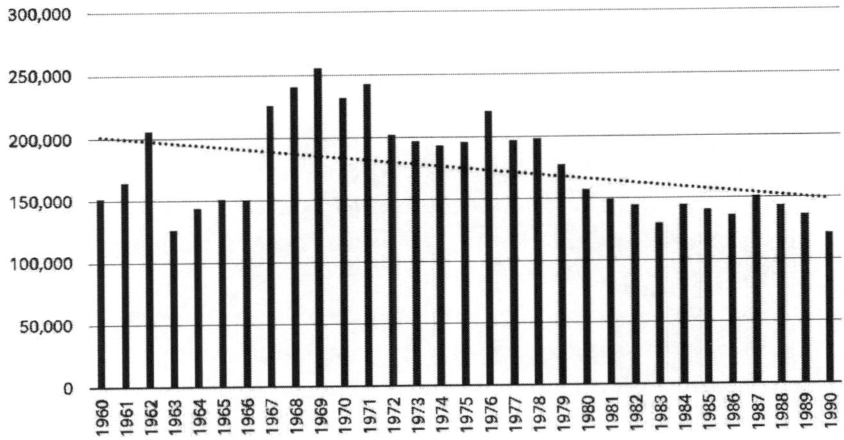

Fort Michilimackinac attendance, 1960 to 1990.
SOURCE: FIGURES COURTESY MACKINAC ISLAND STATE COMMISSION.

in Mackinaw City would be another oil well, financially self-supporting in its first year of operation, and eventually making a profit.[30]

Attendance at Fort Michilimackinac confirmed Woodfill's fears, rather than his hopes. The commission estimated in 1960 that 225,000 paying visitors were needed to make Fort Michilimackinac self-funded. In 1960 only 151,744 tourists came.[31] It would be 1967 before 225,000 people visited Fort Michilimackinac.[32] Over the thirty years between 1960 and 1990, attendance, on average, annually declined. Fewer people visited the fort in 1990 than in 1960.[33]

Despite the disappointing financial figures from the Fort Michilimackinac site, the commission continued to invest in new Mackinaw City projects. In addition to ongoing work at Fort Michilimackinac, projects in or near Mackinaw City included the development of a maritime history center adjacent to Fort Michilimackinac, the construction of a full-scale replica of an eighteenth-century sailing vessel that had originally been built at the straits, and the development of a new park site approximately four miles east of Mackinaw City.

In late 1967, the commission voted to close the remaining campsites at the campground it had operated in Mackinaw City since the 1920s and use the land for a maritime museum, to be erected close to the decommissioned Mackinaw City lighthouse, over which the commission had taken control.[34] The museum was constructed, but visitor attendance did not generate sufficient revenue to cover costs and it was closed in the 1980s, although the commission continued to keep the lighthouse open to the public.[35]

Reconstructing the sailing vessel *Welcome* was another project undertaken by the commission to create a maritime attraction in Mackinaw City. The first *Welcome* had been a fifty-five-foot-long cargo boat, built in 1775 by the fur trader John Askin on or near the ground where the commission had placed its maritime museum. Construction on a full-size replica of the ship began in 1972. Once completed, the ship proved a popular tourist attraction, but not a profitable one. Each year, approximately 865,000 people viewed the ship at the dock, for free, but of that very large number, only approximately 36,600 visitors paid to come aboard. Annual ticket sales did not pay for the ship's annual operating expenses. In 1980, ticket sales were ended, and after sitting at the dock for years

The *Welcome* underway. COURTESY MACKINAC ISLAND STATE PARK COMMISSION.

growing increasingly unseaworthy, in 1989, the commission disposed of the *Welcome* by essentially giving it to the Maritime Heritage Alliance, which relocated it to the alliance's home port of Traverse City, Michigan.[36]

The commission's next venture on the south shore of the straits proved more successful. In the early 1970s, the park staff began to explore developing a new park about four miles east of Mackinaw City. Historic Mill Creek Park was located where the Campbell and Dousman families had lived in the late eighteenth and early nineteenth centuries. Their land housed a number of enterprises, including a sawmill, a grist mill, and a farm. The commission obtained the property in 1975 and opened the new park in 1984.[37]

Initial attendance at the new park substantially exceeded that projected in planning documents. But the planning documents assumed attendance would increase over time. Instead, attendance declined. The first year of operation, Mill Creek drew 74,548 visitors. In 1990, that number had declined to 57,349.[38] Mill Creek would become a permanent part of the commission's venue of attractions, but the number of paid visitors would be problematic. For all their efforts at Fort Michilimackinac, attempts to develop a maritime attraction, and efforts at Mill Creek, they had not clearly established Mackinaw City as a self-supporting tourist destination.

The Mackinac Bridge Authority (MBA) also supported tourism in the region in a way that particularly benefited Mackinaw City. The Mackinac Bridge was not designed for pedestrian use, but as soon as it opened, the MBA began to receive requests from people who wanted to walk across the bridge. Larry Rubin, executive director of the MBA thought that most of the requests were "zany things" but one request, from the International Walker's Association (IWA), struck him as interesting.[39] On June 25, 1959, sixty people, all members of the IWA, gathered in Mackinaw City and walked to St. Ignace. The event was a success and, with the hope of helping the tourist trade, Rubin convinced the Bridge Authority to authorize a second Mackinac Bridge walk on Labor Day 1959. About 250 people showed up. It was a modest beginning for an event that would become very popular. In 1989, sixty-nine thousand people walked from St. Ignace to Mackinaw City on Labor Day.[40]

Leaders in both Mackinac City and St. Ignace also worked to develop tourism at the straits by creating tourist activities that expanded on the projects of the park commission and the Bridge Authority or represented new ideas to bring visitors to the area. Just as they had in 1933, Mackinaw City merchants used the fort as a centerpiece for a weekend festival. Although the 1933 reenactment had drawn a substantial audience, at some point, the event stopped being held and apparently was largely forgotten.[41] "The Re-Enactment of the Massacre at Fort Michilimackinac" was reborn in 1963 when, over the Memorial Day weekend, Mackinaw City residents recreated the events of June 2, 1763, when a group of Native American warriors captured the fort.[42]

One factor that may have led to the modern version of the reenactment was that a significant number of white hobbyists had begun to take an interest in Native life. Some had gone so far as to attend summer gatherings dressed up as Native Americans. This had become so popular that a summer circuit, known as the "pow-wow highway," had developed in the Midwest. The highway included annual meetings in Ann Arbor, Indianapolis, and Wisconsin. These gatherings and others were described in a publication dedicated to them, *Powwow Trails*, that began publication in 1964.[43] There was an interest in this kind event that went far beyond the straits area.

The 1963, Mackinaw City event involved about seventy-five reenactors who performed for several hundred visitors.[44] In 1967, the reenactment involved about 250 volunteers and was performed in an area with seating for 2,500. The event grew and added associated activities, such as a muzzle loaders festival and Native dances.[45] In 1969, five to seven thousand people witnessed the reenactment.[46] In 1970, the NBC television network gave the event national coverage.[47] In 1971, because of protests by Native Americans who objected to, among other things, the word massacre in the event's title, it was renamed the Pageant of Fort Michilimackinac.[48]

If the Memorial Day weekend pageant began the summer season, the merchants of Mackinac City generated business at the end of the season from the throngs of people taking part in the annual Mackinac Bridge walk on the Labor Day weekend. The Mackinaw City Chamber of Commerce began to issue "certificates" "to prove that he [the walker] had completed the long journey."[49] By 1966, each free souvenir certificate had printed on it unique lucky numbers. Most Mackinaw City merchants posted a few lucky numbers their store's window. The person whose certificate bore the lucky number posted in the window won a gift from the store. Merchants optimistically assumed that after completing the five mile bridge walk people would take a few more steps into Mackinaw City's business district to see if they had won a prize and, the shop owners hoped, buy a few things.[50] That is what happened, and store owners capitalized on the foot traffic to sell two must-have souvenirs: an "I walked the Bridge" shoulder patch, first sold in approximately 1970, and an "I Walked the Bridge" T-shirt."[51]

In addition to the Memorial Day pageant and Labor Day bridge walk events, Mackinaw City tried to expand the number of visitors coming to the community with other new seasonal activities. These included organizing an annual fudge festival that began in 1980 and an annual Iron Workers festival that began in 1984. The community also tried to attract more bird-watchers throughout the season. To coordinate events and outreach, Mackinaw City hired its first full-time director of the local tourist bureau in 1983.[52]

If Mackinaw City was working hard to develop itself as an independent tourist destination, it also continued to be a transfer point for guests bound for Mackinac Island. In 1960, 376,423 people had ridden ferries in Mackinaw City.[53] In 1989, that number had grown to 871,143.[54] Just as in the old days of the railroad, these people were mostly passing through, but some paused long enough to buy a snack, eat a meal, or purchase a souvenir. This increase in passengers also substantially increased the income of the city's ferry boat companies.

The ferry business from Mackinaw City to the island had begun in 1886 when George Arnold began scheduled passenger service from Mackinaw City to Mackinac Island.[55] The Arnold ferries long dominated the trade, but after World War II, they began to receive serious competition.[56] What proved the most challenging came from speedboat operators. By 1937, the White Star Rapid Transit Company was running a scheduled speedboat service with eight daily sailings from Mackinaw City to the island.[57]

Rivet-eye bridge tour, ca. 1956.
COURTESY JOHN MARSH POSTCARD COLLECTION, MACKINAW AREA PUBLIC LIBRARY.

The speedboats were much faster but much smaller than the Arnold ferries. The ticket was likely more expensive, but they served people not wanting to wait for the next scheduled sailing of an Arnold ferry, or those who simply wanted to get to the island more quickly.[58] During the construction of the Mackinac Bridge, speedboat operators found a new way to make money: taking guests out to the bridge construction site for a "rivet-eye view" of what was happening.[59]

Among the most successful of the post World War II speedboat operators was the Shepler family of Mackinaw City. William H. "Cap" Shepler was in charge. After World War II, Cap accepted a job piloting an Arnold ferry. But Cap dreamed of owning a speedboat. To make his dream come true, he bought a twenty-eight-foot mahogany and chrome "fixer upper," which he rebuilt over one winter. The boat made money, but to really compete with the Arnold line, he needed a ferry of about sixty feet in length that could carry over one hundred passengers and cruise at about thirty miles per hour. His first effort to build such a boat ended in failure. In 1969, Cap Shepler had his second new boat built, the *Welcome*. Cap

Shepler personally sailed the ship into the straits on its maiden voyage, racing under the Mackinac Bridge at full speed. Cap declared it "a Godamn speedboat." It was a speed boat and, more importantly, a commercially successful vessel that led to a small fleet of additional ships.[60]

Shepler's success, as well as that of other ferry lines, was based on the rapid growth in the number of visitors going to Mackinac Island. In 1958, 258,526 passengers had been taken to or returned from the island from a dock in Mackinaw City. By 1964, that number had grown to 667,794.[61] The growth in passenger traffic to the island grew the Sheplers' business, and the Sheplers, in turn, encouraged new businesses in the community. In 1974, for example, Bill Shepler, Cap's son, persuaded Audrey and Edgar Jaggi to relocate to Mackinaw City to purchase a local restaurant, which eventually was named Audie's.[62] People believed that the straits tourist industry could make for a good life.

Growth was also seen in Mackinaw City's hotels, although the accommodation business was highly cyclical. In 1958, the community had forty-seven hotels. The number of hotels in the city in 1986 was fifty-six, but these numbers conceal a roller coaster ride. Hotels first peaked in 1967 when the city had fifty-eight hotels, but then fell to forty-five in 1973. But for every bust there is a boom, and in 1982, the community had sixty-two hotels. By 1984 only forty-eight were still open.[63]

Counting the number of hotel rooms available in Mackinaw City each year is impossible, but in 1984, 1,217 units were available each night in Mackinaw City, a modest increase over the 1,000 mentioned in the early 1960s. The extra two hundred rooms were likely accounted for by the closing of small "mom and pop" establishments and their replacement by larger motels, some affiliated with national motel brands. Although locally operated motels with names like the Bridgeview, Parkside, Riviera, and Vin-Del remained, new motels were found in town, affiliated with national chains such as Best Western, Super 8, and Ramada.[64] The three decades after the opening of the Mackinac Bridge there had been a modest increase in the availability of overnight accommodations in Mackinaw City, but with great variability from year to year and a cycle of booms and busts.

St. Ignace's leaders also tried to build up their tourist economy. One way was to continue its tradition of annually celebrating the community's

Black Gown Tree, Pere Marquette Pageant, St. Ignace, ca. 1967–1971.
COURTESY MICHILIMACKINAC HISTORICAL SOCIETY, ST. IGNACE.

link to the Catholic missionaries who had come to North America in the seventeenth and eighteenth centuries by commemorating Father Jacques Marquette. The exact nature of the commemorative program varied from year to year, but the event regularly drew a large audience. In 1957, for example, two thousand people attended the celebration.[65]

In 1966, the Pere Marquette Pageant was reimagined as a musical drama featuring a cast of about one hundred. Recognizing talent when they saw it, the organizers from St. Ignace hired to run their Labor Day pageant the man who put on Mackinaw City's Memorial Day Fort Michilimackinac Pageant.[66] The pageant was also linked to a small museum honoring Father Marquette. The museum had originally been constructed in 1837 as a local mission church. It had been replaced by a new church in 1904, and in 1954, the old building was moved by the Knights of Columbus to a location adjacent to St. Ignace's Marquette Park. After being moved, the church building was repurposed as a museum.[67] In 1976, as part of the US Bicentennial, the community successfully undertook a campaign for a greatly enhanced tribute to Father Marquette.[68] In 1980, the Father Marquette Museum was opened.[69]

The most successful new effort to attract tourists to St. Ignace, however, celebrated not local history but rather the American automobile. It was developed by local banker Ed Reavie, who in 1975 suggested holding a collectible car show. Reavie, who was well known in car circles, called friends to see if they would bring their cars to St. Ignace. Many agreed, if Reavie would pay gas money and the toll to cross the Mackinac Bridge. Reavie didn't have the cash, so he went to the city, which much to his surprise, agreed to pay five dollars to every person who officially entered a car in the show.[70] Reavie's goals for the show had been modest. He hoped that perhaps twenty-five or thirty cars would be on display.[71]

Reavie asked around to find out what was the summer's slowest business weekend. It was the last weekend in June, so he picked that weekend for the show.[72] On June 25, 1976, the first St. Ignace car show was held, with 136 cars viewed by approximately 5,500 people.[73] The response was "unexpected and overwhelmingly positive," and the show became a much anticipated annual event, eventually expanding from a Saturday afternoon to a three day festival.[74]

In 1978, Reavie conceived of bringing a "guest of honor" to the show, someone who was a celebrity in the auto world. The idea was a hit. One of the most well received guests of honor was Carroll Shelby. Shelby's visit also was responsible for one of Reavie's favorite stories. While Reavie was driving Shelby to an event, Shelby announced he needed a beer. Reavie obliged and stopped at a local bar. Concerned about how long Shelby had been inside the bar, Reavie went to look for him. He found Shelby, happily answering questions posed by fans. As Shelby made his goodbyes, he also made his amends to Reavie by buying two bottles of beer, one for himself and one for Reavie.

As the two men hoisted their open beers in salute to one another Reavie pulled onto the road. He was promptly pulled over by a state police car. Driving in Michigan with an open bottle of beer was illegal. Reavie was dreading his ticket and the bad press it would cause. To his surprise, the police ignored him and, more importantly, ignored the beer. The officers wanted to talk to Shelby. One said, "Mr. Shelby, welcome to St. Ignace," and asked for an autograph. Shelby obliged and the officers walked back to their car, without issuing a ticket. As the police pulled away, Shelby quipped, "Nice town you have here."[75]

Reavie again expanded the auto show when he invented Cruise Night. Cruising to a drive-in restaurant with the car radio blaring had been a regular part of Reavie's teen years. Cruise night relived that experience, with a long line of vintage cars taking to the road and a DJ at the destination drive-in playing '50s hits. If a DJ spinning records was a good idea, a live music show was an even better idea. Reavie began producing "The Legends of Rock and Roll," a live show held in the local high school auditorium. The legends who found their way to St. Ignace included national acts such as Gary Lewis and the Playboys, Lesley Gore, and the Shirelles.[76]

As the number of cars being displayed grew, someone suggested a parade. It was a good idea with one problem: Reavie hated parades. But a colleague, Merv Wyss, loved parades and proved a master at organizing them. The Down Memory Lane Parade began in 1982 and soon became a crowd-pleasing fixture.[77] One of the most memorable vehicles in the parade was the Batmobile, featured in the Batman TV series that aired from 1966 to 1968. It was brought to town by guest of honor George Barris, the "King of the Kustomizers" and the man who had constructed the car for the television show.[78]

By 1990, three thousand cars were displayed during the auto show weekend, enjoyed by more than one hundred thousand visitors. Community leaders recognized the value of the auto show. The St. Ignace Downtown Development Authority director could not say enough nice things about it. "I watched our little town go from shooting a cannon down Main Street and not hitting anyone to all those incredible activities. It all began with the car show."[79] Indeed the 100,000 people who came to St.

Ignace for the car show in 1990 rivaled the 120,928 who had visited Fort Michilimackinac in the same year. This was tourism on a grand scale, but it put St. Ignace on the map for a weekend in June rather than as a year-round tourist destination. Huge as it was, it was not enough to change the perception of St. Ignace as a place that, for most of the year, people visited to go to Mackinac Island.

While the efforts of Ed Reavie in St. Ignace and members of the Mackinaw City business community were significant in attracting visitors to each community, their accomplishments were balanced against disputes that adversely affected the tourist industry. These disagreements are best documented in Mackinaw City. The Mackinac Island State Park Commission's efforts to rebuild Fort Michilimackinac in Mackinaw was an early flashpoint. Before work on Fort Michilimackinac even began, commission chair W. Stewart Woodfill alienated one of Mackinaw City's major community organizations, the Lions Club. During the construction of Mackinac Bridge, the club was given permission by the Park Commission to place pay-for view binoculars on park property. Tourists inserted a few coins to get a close-up look at construction, and the money raised funded Lions Club projects. Woodfill, citing requirements that all park income must now be directed toward bond payments, had the Lions Club binoculars removed and replaced by ones whose coins would go into the commission's treasury. The Lions were furious.[80]

It was a bad start to what would become an increasingly stormy relationship. The largest explosion occurred because of the revenue shortfall at Fort Michilimackinac. The park was losing money and surveys revealed that of those who came to use the park, about one-third paid admission to the fort, another third paid to camp at the campsite, and the remainder visited the park without paying anything, enjoying a picnic or simply taking in the view. Losing money was not in Woodfill's nature, and he vented angrily about the presence of those who paid nothing.

> I see no point to our giving a free ride to people who simply want to enter the park and view the bridge, picnic, use our toilets—and refuse to pay 50 cents to inspect the fort. To hell with them. They only litter up our place, take up our parking space, and aren't worth the powder to blow them up.[81]

If the freeloaders in Mackinaw City weren't worth the powder to blow them up, extracting an admission fee from them was worth the price of fencing in the entirety of the commission's Mackinaw City property.[82] As soon as the idea of fencing the entire property and charging everyone entering an admission fee became known, vehement opposition to the idea developed. Instead of rethinking his plan because of this opposition, Woodfill decided to test what would happen if he actually made everyone entering the park pay a 50-cent admission fee.[83] What happened was a explosion of ill will.

Although the date of the test was supposedly a secret, it seems everyone knew about it. Before the park opened on the morning of the test, a crowd had gathered at the park entrance. On one side of the street were a group of town residents, including the mayor. On the other side of the street stood Woodfill, Petersen, and the commission's sole uniformed police officer. A line of cars had formed, waiting for the park to open. Just before the park gate was to go up, Woodfill instructed the commission's police officer to step forward and ask the driver of the first car for fifty cents. As the officer entered the street, the village's mayor immediately ordered the officer out of the street. Looking directly at Woodfill, the mayor said he was going to get a court injunction to stop the test. In response, Woodfill asked the officer for his hat and coat, put them on, and he approached the first car, spoke with the driver, and then worked his way down the line, while the mayor looked on.[84]

If the mayor was left without words that morning, he and the village government quickly found their voice. The Village Council, meeting a few days later, passed several resolutions that made various demands of the park commission. The most important was that picnickers continue to have free access to the park. Another resolution demanded that park campsites "must not be abandoned ever."[85]

The village government had good reason to be concerned about the campground. Since the 1920s, campers at the site had represented a steady source of income for local merchants. If it were relocated away from town, local businesses would suffer. Renovating Fort Michilimackinac had already required relocation of the campground from the west side of the newly constructed Mackinac Bridge to the bridge's east side.[86]

This may have fueled rumors that eventually the campground would be eliminated. Woodfill, however, was not considering this. The campground generated approximately $16,000 each year in revenue, income his fellow commissioners very much wanted to continue.[87] But the days of the campground were numbered. In 1968, a new park master plan called for entirely relocating the campground outside of Mackinaw City, a project eventually completed in cooperation with the Michigan Department of Natural Resources.[88]

In 1960, however, the fight was over fencing, and it was a fight Woodfill lost. To explain what happened, Woodfill said, "Those people in Mackinaw City have a deep proprietary interest in that Park and its operation, and they are the craziest people I have ever been mixed up with."[89] Over the long term, the relationship between the commission and Mackinaw City's leaders was repaired. After Woodfill resigned as chair of the park commission, new commission leaders established peace with the Mackinaw City government.[90] During the same 1982 legislative investigation where St. Ignace's city manager criticized the commission, the mayor of Mackinaw City had kind words for the body. When a committee member pushed the mayor for a complaint, the mayor returned to the old issue of the fence, saying that he was concerned that the beach within the fort's fenced area was not open to "through traffic." But the mayor quickly added that his concern "was small compared to the benefits the community obtains from the park programs."[91]

But repairing the relationship between Mackinaw City's leaders and the commission was not enough. It was not enough because Mackinaw City itself was often unthoughtful or divided over how to develop the community's tourist industry. One example is the dock built on the Mackinaw City waterfront by Merritt-Chapman & Scott, one of the Mackinac Bridge's largest subcontractors. The company used the dock to load barges carrying cement to the bridge construction site. When the bridge was completed in 1957, the company had no use for the dock and donated it to the village. It was an opportunity to invest in a community showpiece, one likely to be seen by the 258,526 ferry boat passengers in 1958, in addition to the many visitors who never boarded a ferry.[92] Instead of a showpiece waterfront park, local government, intent on gaining new revenue, leased the site to the Shepler firm, which used the dock for its ferry boat business.[93]

In 1985, Mackinaw City created a Downtown Development Authority (DDA).[94] In early 1987 the DDA released a proposal that it claimed would revolutionize tourism in the town, transforming it from being simply the "parking lot for Mackinac Island." Key to the plan was a historic walkway. Said the DDA, "History sells and people will travel to learn about it, they will spend money to enrich their learning experience."[95] Although there was some criticism of the plan over how the historical walkway was conceptualized, very quickly the issue became the tax scheme to be used to finance the proposal.[96] Arguments over taxes were long and loud and eventually led to the proposal being abandoned.

There were other community controversies in the 1980s linked to tourism. These included an unsuccessful effort to create a unified community architectural plan such as the one on Mackinac Island, unhappiness over how the local tourist bureau was operated, and efforts, ultimately unsuccessful, to encourage light industry in the community despite possible negative consequences on tourism.[97] The local newspaper, which had maintained a neutral position while the DDA controversy raged, wrote after the fact, "In pondering on the future of the community, we have noted the voice of the people. You have spoken and made your words clear. You do not want to enter into anything that may cost more money."[98]

The result of these controversies—more properly, the failure to agree upon how to develop tourism within the community and the unwillingness to pay for the infrastructure that would attract tourists—left Mackinaw City unable to claim that it was a tourist destination in its own right. It remained viewed within the state, as the DDA had labeled it in 1985, as Mackinac Island's parking lot.

St. Ignace also often acted in unthoughtful ways regarding tourism or in ways that did not prioritize tourism over other opportunities. At the height of speculation regarding how the Mackinac Bridge would stimulate tourism, in 1958, the St. Ignace Civic League sold its waterfront park to a person who planned to build a gas station on the land. The Civic League planned to use the money from the sale to rehabilitate the clubhouse at the city's golf course. This was not a proposal hostile to tourism, but the golf course was not prime waterfront property that would create a first impression on visitors. The Civic League was eventually embarrassed into reversing itself.[99] It was a good decision, considering that in 1958, 385,856 people traveled through the St. Ignace harbor.[100] With so many people riding ferries, the appearance of the town's waterfront mattered more to tourism than the condition of the local golf course's clubhouse.

In subsequent years, St. Ignace discussed creating a unified architectural theme for its downtown. The community seemed to have developed a plan, modeled on what had been accomplished in Gaylord, Michigan. Gaylord had adopted a unified architectural theme in the hopes of rebranding itself as an alpine village. In Gaylord, the goal was to keep costs low, thus work was done with modern materials, and not with fully accurate historical detail but rather in a "reminiscent contemporary style." However, in St. Ignace, the plan did not lead to concrete results.[101] When new initiatives did succeed, the results often proved disappointing. The Marquette National Memorial, for example, attracted only a modest number of visitors each year. Between 1985 and 1990, attendance at the Marquette National Memorial averaged approximately 18,100 visitors annually, compared to the 138,000 people on average who annually visited Fort Michilimackinac in Mackinaw City during the same years.[102]

Even something as successful as the St. Ignace auto show generated complaints. Some people were annoyed by the number of street closings that inevitably were needed to accommodate the large number of vehicles on display and the crowd they attracted.[103] Crowd control caused other problems. Far more troublesome was that by the late 1980s, as the local newspaper editor expressed the concern, "it [the auto show] was one big drunken party."[104] Some local bands refused to march in the car parade due to problems with drunks.[105] Reavie himself wrote, "The drunkenness and rowdy behavior hurt this show. It hurt my relationship with my sponsors and even my supporters with the city."[106] Eventually a strictly enforced zero tolerance policy for alcohol or any other law violations brought back both sponsors and the family character of the event. When enforcing the law, the police did not play favorites (unless Carroll Shelby was in the car). One year, Reavie parked in a no parking zone. When he came back his car was gone, ticketed and towed by the police.[107]

Other problems demonstrated the shortsightedness of some in the city's tourist industry. When Bootsies Drive-In, the original destination for Cruise Night, went out of business, Reavie looked for a new cruise destination. All of the other drive-in restaurants in town turned him down. They claimed they were already too busy that night to serve as the event's destination. An incredulous Reavie pointed out they were all too busy because of the traffic the cruise was creating and to make the cruise happen he needed a destination. Their shortsightedness nearly ended the event from which all of them were handsomely profiting. Eventually a local restaurant on the highway near where the drive-ins were located agreed to be the destination, although it was not a drive-in.[108]

Although both Mackinaw City and St. Ignace had long history of people driving to spend the night in campgrounds or other comparatively inexpensive overnight accommodations, the perception remained that they were mere annexes of the island. As *Detroit Free Press* columnist Bob Talbert wrote in 1976, "Without the Island and the fudge, St. Ignace wouldn't exist."[109] A decade later, Bill Shepler, Cap's son, inadvertently applied Talbert's opinion to Mackinaw City. In 1987, he rebuilt the automobile entrance to the family's Mackinaw City dock, one of the most prominent locations in town. A key feature of the project was a handsome arch through which cars passed. On the arch were the words "The Gateway to Mackinac Island."[110]

The writing of Talbert and the words on the arch leading to the Shepler's dock were reminiscent of how people viewed the shoreline a hundred years earlier, as the place the train stopped and passengers caught the ferry to Mackinac Island. Perhaps Talbert and Shepler should have written "The more things change, the more they remain the same."

Epilogue

After spending much time discussing tourism at the Straits of Mackinac, several questions merit additional thoughts.

- What is the relationship between the Native American community and the tourist industry?
- What is the significance of conflict within the tourism industry?
- Why was there resistance within the straits communities to embracing tourism as the region's primary industry?

The Relationship Between the Native American Community and the Tourist Industry

There are two issues that come to the fore regarding the Native American community and the tourist industry. One involves how people who came to Mackinac Island with the intent of interacting with the Native community did so. The other is how the business relationship between white merchants and the Native community operated.

Even a quick reading of the pre–Civil War literature makes it clear that many travelers came to Mackinac Island to see and speak with Native Americans. The writing of both Anna Brownell Jameson and Margaret Fuller makes this point. They came to observe, communicate, and learn.[1] Because Mackinac Island was the location where the federal government made treaty-required annuity payments to Indigenous people from throughout the Great Lakes region, the island was a gateway to talk to both Native Americans who lived there and Indigenous people from throughout the Great Lakes region. The tradition of going to Mackinac Island to see and speak with Native Americans endured long into the twentieth century. For example, the 1941 Old Mackinac Lives Again celebration promised visitors the opportunity to interact with Chippewa and Odawa tribal members. Native Americans wearing traditional costumes would be present. Native American craftspeople would be demonstrating their traditional skills. There would be a Native American village erected on the island's beach. Tourists could even try their hand at making Indigenous craft items, with the help of a Native American tutor.[2]

This interaction between the Indigenous people of the Great Lakes and visitors to Mackinac Island raises an interesting problem: how to interpret what was learned. Did the information received come from a Native American who was a life-long island resident or from someone from a different tribal community hundreds of miles distant from the island? Well into the twentieth century, as the promotion for Old Mackinac Lives Again makes clear, a visitor could be talking to an individual from different tribal communities. Yet the literature created from these conversations simply describes what "Indians" said and did. To better interpret what visitors who spoke with Native Americans on Mackinac Island learned involves the challenging process of sorting out who the teacher was, where the teacher came from, and what values and practices were common in the teacher's community.

Many visitors to Mackinac Island never saw a Native American but did take home a souvenir made by a Native craftsperson. The trade in Native-made goods required merchants on the island, and later in St. Ignace and Mackinaw City, to interact with Native American craftspeople

Native Americans on Mackinac Island, 1903. Francis Benjamin Johnston, photographer. COURTESY THE LIBRARY OF CONGRESS.

who were manufacturing the goods. On Mackinac Island, scholars have documented a significant and well-organized Native community that existed in the nineteenth century with well-respected leaders.[3] It is not clear, however, what role, if any, these leaders played in the tourist industry. There is little documentation of how merchants obtained the Native goods they sold. But that the marketing of Native crafts was an important part of the tourist trade is seen in the ads of nineteenth-century island merchants and the hopes of the organizers of events such as Old Mackinac Lives Again. The people who organized Old Mackinac Lives Again were particularly insistent on the scope and quality of their Native exhibits, saying they wanted their display of Anishinaabe crafts to rival shows of Navajo and Hopi exhibitions that occurred in the American Southwest.[4]

One of the few interactions between the Native community and the island's merchants that can be documented suggests that well into the twentieth century, the merchants were anxious to preserve their role as middlemen between the tourists and the Native community. In 1966, the North American Indian Association opened a store on the island selling "authentic Indian handicraft and beadwork."[5] The store was quickly closed by the city, allegedly because the building it was in did not meet code. Some believed, however, that the store was closed because the "council was pressured by store owners selling imported beadwork."[6] After what was likely a tense few days, the Indian Association was allowed to resume selling their material, but at an outdoor stand near the Arnold Ferry Dock rather than from inside a store. The Indian Association felt they had been treated badly, saying, "We've been relegated to a second-class position where many tourists never see us, and if they do they're in too great a rush to get on or off the boats to buy anything."[7]

St. Ignace and Mackinaw City also had Native American residents who, like the Native American community on the island, participated in the tourist trade. One of the best documented interactions between a merchant and a Native community happened in St. Ignace.

When, in 1925, Clarence C. Eby opened Indian Village, he took as a business partner Duncan Bajbedauscen, who was a Native American. There is no parallel to this in either Mackinaw City or Mackinac Island. The business showcased Native people engaged in craftwork, and Bajbedauscen's position as a partner in the business likely served to enhance the authenticity of what a visitor saw.[8]

Eby's attituded toward Native Americans was different from the widespread nineteenth-century narrative regarding Native-white interaction that many of his era would continue to believe—the inevitable conquest of the uncivilized Native by the superior white man. Eby would reflect years later,

> As the once mighty tribes gave way before the overwhelming conquests of war and subsequent exploitation, they retained securely for themselves exclusively the secrets and knowledge of unique arts and crafts. . . . Today these arts are becoming more and more recognized at their true value, . . . and while it is impossible to reproduce all of the finest of Indian wares, yet many genuine Indian products can be and are produced which put to shame the shoddy imitations which have so often crept into the market.[9]

Eby's choice of words to describe Native American history after the arrival of Europeans to North America, "conquest" and "exploitation," suggests a fundamental reinterpretation of American history and an equally fundamental change in the attitude of some local merchants toward their Native partners.

Harry Teysen, who opened a restaurant in Mackinaw City in 1926, had a deep interest in Native life. As a result, he developed a large, personal collection on the topic. Eventually the collection was housed in his restaurant as a Native American museum. The museum was a roadside tourist attraction, but it also seems to have had the goal of educating tourists to a better understanding of Native culture. His son, Ken Teysen, who relocated and rebuilt the museum in the upper floor of the family's restaurant, employed professional museum staff to display and interpret the collection. Ken Teysen was also clear in saying that he rejected the stereotype that Natives were poor, ignorant, and savage.[10]

The Teysen family and Eby seem to have had particularly enlightened attitudes toward their Native American neighbors. But they were not the only commercial sites that recognized the appeal of Native American themes to tourists. For example, when the newly built Fort Restaurant & Gift Shop, located in Mackinaw City, opened in 1963, the decor was designed to blend with the nearby Fort Michilimackinac and the waitstaff was attired in Native American dress.[11]

The belief in the 1960s and 1970s that tourists were interested in Native history and culture led Mackinaw City merchants to reenact the events of June 2, 1763, when Native American warriors overran the British garrison at Fort Michilimackinac. The event had first been staged in 1933, as part of the dedication of the restored Fort Michilimackinac stockade. In 1933, to improve the "authenticity" of the reenactment, a serious effort was made to recruit Native Americans to participate in the event. They set up camp inside the stockade and entertained with dancing, basket-making, stories, and by just appearing in Native costume.[12] Sixty-five Native Americans participated in the event.[13]

Much of what transpired at the reenactments reinforced stereotypical views about Native Americans. One example was the tale of Chief Thundercloud. The day before the first Fort Michilimackinac reenactment was to happen, the weather was poor and there was concern that the event might be rained out. Hearing this, Chief Thundercloud, who was smoking a cigar, supposedly said, "Me fix it." The Chief blew smoke, danced a bit,

Chief Joe Shomin outside the original reconstruction of Fort Michilimackinac.
Courtesy Clarke Historical Library, Central Michigan University.

and then buried some of his cigar ashes. The next day the sky was clear and a local legend regarding the efficacy of a stereotypical Native dance was created.[14] Chief Thundercloud became part of the lore of Mackinaw City's tourist industry.

For reasons that are unknown, the reenactment that began in 1933 ended. Native dancers continued, however, to be a tourist attraction, including a troupe of fifty dancers recruited from the Harbor Springs area who performed at the Mackinac Bridge dedication.[15] Dance demonstrations, however, lacked the appeal to draw a large number of tourists. Thus, during the 1963 Memorial Day weekend, the idea of reenacting the events of June 1763 was born anew. Like its predecessor thirty years earlier, it proved a very successful tourist attraction.[16] But it also became a window through which to view the emerging interaction between straits-area tourism and the Native American community.[17] In 1971, a Detroit newspaper reported that some members of the Native American community were criticizing the event. A Native spokesperson said, "The white merchants in Mackinaw dress up a lot of white people in costumes and war paint and put on a show that does nothing but serve the business interest of the white community." The "Hollywood style" restaging of the battle portrayed Natives in a "violent, racist atmosphere." Some members of the Native Community asked that the event be canceled. Others were not opposed to the event's continuation but wanted to add a more accurate representation of Native life. "The real tragedy is that there is no museum or cultural center that receives the attention and publicity of this pageant, and which emphasized the peaceful side of Indian Life."[18]

Some members of the Mackinaw City business community dismissed the complaints. Said one local gift shop owner, "They [Native Americans] don't want us to do anything that would portray the Indian in an objectionable manner. Well, we've taken the story of the battle straight from the history books, and one thing we're not going to start doing is rewriting history." Likely history books had been consulted, but the history books had been written from the perspective of the white community. As for using the event to fund a museum, another local merchant said, "We can't give them any profits, because we don't make any."[19] Admission to the reenactment was free, thus technically no money was made. However, the event's purpose was to increase the number of tourists visiting Mackinaw

City with the hope of increasing revenue throughout the town over the Memorial Day weekend.

Eugene Petersen had warned in 1963 that the reenactment could cause controversy. The park commission had been asked to contribute $1,000 to help defer expenses incurred by the event, and Petersen had said that if the program was not "dignified and authentic" there might be a problem.[20] Cochairs of the 1971 event, Ken Teysen and Raymond Desy, took a page from Petersen's comments, arguing that the script was authentic. They said they "considered the script, written by a recognized historian, to be historically accurate." Realizing, however, that the historical accuracy was not at the heart of the controversy, they added, "We have never intended to defame the ancestors of our Indian friends." To make amends, script changes were made and the event was renamed. When challenged as to what to call it, a Native spokesperson said "military victory." "The Re-enactment of the Native American Military Victory at Fort Michilimackinac" was a bit too much of a change; the event was henceforth known by the perhaps purposely vague title of the "Fort Michilimackinac Pageant."[21]

That a compromise was struck may have happened, in part, because one of the pageant's principal organizers was Ken Teysen. Teysen had a deeper understanding of Native American history than many of his fellow business owners, who cited history but likely did not know it well. Teysen also was wise enough not to talk about profit, unlike other members of the business community who were disingenuous regarding the money they hoped the event would bring them. His understanding seems to have calmed the situation. But it was not a calm that created the basis of a more lasting understanding, an ongoing dialog between the town's business community and the Native American community. A memo written by the Pageant Committee described what had happened this way: "Through negotiation [we] warded off a possible militant Indian uprising."[22] Trouble had been averted, but an opportunity for a partnership was not explored.

Some members of the Native American community may have been aware that they were not being incorporated into the event's decision-making process, and thus chose to picket the 1971 pageant.[23] In 1972, it became even more obvious that Native input was not being sought when a Native

American dance competition was held. The event drew participants from five states, but it was organized by a non-Native group affiliated with the Boy Scouts of America. The event may actually have been held to attract travelers on the pow-wow highway, as it also appears that most if not all of the dancers who performed were not Natives.[24] This was not the first time that non-Natives had been featured in a pageant-sponsored Native dance exhibition. For example, in 1969, "Authentic Indian dances" were performed by the Chief Wawatam's Tribal Dancers, who were students enrolled in the Mackinaw City High School.[25]

In the 1970s, St. Ignace's political leadership also showed increased sensitivity to the Native American community. A particularly significant example of this occurred in 1972 when the St. Ignace chapter of the Indians of North America protested to the city council regarding the impropriety of a privately owned "Indian burial ground" that was displaying the skeletal remains of Native Americans, charging tourists fifty cents to see them. The remains were likely genuine. The graves of at least fifty-two individuals that an archaeologist had documented to be at least three hundred years old, had been discovered in 1958 while excavation work was being done on private property. Eventually the human remains became an advertised attraction at a local campground. The campground ad that mentioned human remains on exhibit placed them next to other featured attractions, such as a picnic area and clean rest rooms.[26] The St. Ignace city council asked the representative of the Indians of North America to meet with the city attorney and help the attorney draw up an ordinance outlawing the exhibition of human remains.[27]

Clearly Native Americans and members of the straits-area tourist industry were engaged with one another. Over time, that engagement grew and likely changed. Documenting that engagement and how it changed is a topic worthy of future inquiry.

The Role of Conflict

Tourism within the straits area generated persistent conflict. People clashed over their concept of tourism and the tourist experience. One of the best documented of these persistent clashes is the interaction between

Carriages on Mackinac Island, ca. 1925.
COURTESY CLARKE HISTORICAL LIBRARY, CENTRAL MICHIGAN UNIVERSITY.

the Mackinac Island State Park Commission and the carriage operators who conducted tours through the park.

The carriage operators had a troubled reputation. In the early years of the twentieth century, a Detroit newspaper described what visitors arriving at the island dock could expect: "Tourists who have visited Mackinac Island in previous seasons . . . will never forget the manner in which they were received by a hundred yelling, fighting hotel runners and carriage drivers as they stepped from their boat."[28]

The carriage operators' official relationship with the Mackinac Island State Park Commission began at the beginning of the twentieth century, when the commission imposed a fee to operate commercial vehicles in the park. The carriage operators bitterly opposed the fee, and filed legal action against it, appealing the matter to the Michigan State Supreme Court, which ruled in favor of the commission.[29] In 1917, the park commission moved on from simply licensing carriage operators to attempting to regulate the trade's business practices. In the commission's opinion, some carriage operators had unreasonably shortened the length of tours they gave and were overcharging tourists. To resolve these concerns, the commission demanded both an "official" tour route and "stipulated" fees. The commission threatened to ban from the park any driver who did not follow the official route or charged more than the stipulated fee.[30]

The commission continued to press the carriage operators to adopt uniform practices and to also create a single organization with which the commission could negotiate, rather than having to deal with many independent operators. The eventual result, in 1922, was the formation of the Mackinac Island Carriage Men's Association.[31] The carriage men, however, wanted something in return for organizing: exclusive rights to operate park tours.[32] In 1924, after about two months of negotiation, an agreement was reached between the carriage men and the commission. The agreement required that all drivers be bonded, drive safely, "be versed in the true history of Mackinac Island and the State Park thereon," and follow the commission's other rules. In return, the association seems to have received exclusive rights to operate tours within the park.[33] But the operators soon decided what they had agreed to was impractical. A year later, a large number of carriage operators attended a commission meeting to express, in the words of the commission minutes, "in an able manner," their objections to the agreement.[34] In 1928, the commission renewed the Carriage Men's Association's exclusive contract for five years.[35] Despite this seemingly tranquil relationship, old issues festered, with the commission, in 1928, expressing concern regarding the history shared by carriage operators with their guests: "The tourist depends entirely on the carriage men who drive them around the Island for a sing-song description of the Island—some of the descriptions are laughable—some are very good."[36]

The decade of the 1930s saw the relationship between the commission and the carriage operators deteriorate. In 1930, fourteen new regulations were imposed on the carriage operators which attempted to regulate their behavior not just in the park but on the city streets and at the city boat dock. Drivers were to be "cleanly, civil, and courteous to the public." Those who were found wanting were threatened with the usual consequence, the loss of their license.[37] In 1931, the commission again complained about the quality of the tour narration given by the carriage operators, citing "ridiculous errors." The commission wanted the operators to create a "standard of information" and, reflecting the 1930 concern over civility and courteousness, a code of ethics. Depression-era financial woes were also leading the commissioners to look at how they might extract more money from the drivers, something that was always a touchy subject.[38]

In 1936, public complaints regarding carriage operators over "(1) rates, (2) chiselers, (3) drivers annoying people, (4) condition of horses and regulations" led the commission to tell the Carriage Association to "clean up their own organization" or the commission would put in place new rules.[39] Cleaning up, however, was easier said than done.[40]

The wartime lull in tourism temporarily ended the complaints by the commission to the carriage operators, but postwar prosperity brought new conflict. In 1948, the commission was again deeply disturbed by the carriage operators' conduct and once again threatened new regulation unless something was done. The twenty-seven owners of the island's fifty-seven carriages licensed to tour through the park responded by forming a new legal entity, Mackinac Island Carriage Tours. Its president, Alex Gillespie, conceded that what had led to the creation of the new body was the operators' past conduct. With considerable understatement, he admitted that, in the old days, "everyone was for himself" and there was "considerable hawking for business" which "annoyed" some visitors.[41]

Contemporary accounts made it appear that the creation of the Mackinac Island Carriage Tours was a voluntary decision. Subsequent tellings of the story made clear it was a much more complicated matter. To obtain what it wanted, the commission used a carrot that had worked in the past. If the operators formed a new organization with rules that were agreeable to the commission, the commission would grant the new organization an exclusive, thirty-year lease to operate in the park. The lease was so generous Commissioner Bill Doyle had to maneuver a bill through the legislature to make it legal.[42]

One would have thought that a thirty-year lease would have created a generation of peace, but the peace lasted a mere two years. In 1950, the commission passed a lengthy set of rules that the carriage operators were expected to follow because "during the current season the rule has been violated in several instances."[43] In 1951, the carriage operators conduct was again before the commission, which threatened that that no new licenses would be issued to the Carriage Men's Association for the 1952 season until satisfactory terms were worked out.[44] This included the old demand that drivers be "more properly schooled in historic information."[45]

In the late 1950s, the park commission, headed by W. Stewart Woodfill, was again concerning itself with the carriage operators. Woodfill wanted uniforms, reasonably accurate commentary that fell in line with the commission's new emphasis on history, an end to the constant pleading for tips, and a new mandatory carriage stop at the Avenue of the Flags, the "back door" entrance he had constructed to Fort Mackinac which allowed visitors to walk down to the fort from their carriage rather than reach it by walking up the long hill that began at the island's dock.[46] Most of this was not new. A resolution requiring the carriage operators to wear uniforms had first been approved by the commission in 1941.[47] Historical accuracy had been a concern as early as 1924.

Tips became a particular point of irritation between the drivers and the commission. The operators asserted that tips had always been part of the business, an integral component of the driver's wage. Although a legitimate point, Woodfill was disturbed by the schemes which some drivers used to obtain tips. One relatively benign gimmick was to have one driver wish another a happy birthday when the two carriages passed, a favor to be returned later in the day when the carriages again passed and the recipient of the first birthday greeting now wished the other driver happy birthday. Neither driver was celebrating a birthday. Another scam involved college tuition. An earnest sounding driver claimed to be having a hard time of it financially but expressed determination to stay in school and "make it," despite everything. A particularly hard sell involved an unauthorized stop in the middle of the woods, where the driver would turn around and address the tourists in the carriage about how poor the pay was and how the company assumed it would be made up by riders tipping the driver.[48]

Woodfill eventually sent Eugene Petersen to lay down his law at a meeting with the carriage operators. Laying down Woodfill's law proved a good way to start a fight. Art Chambers, president of the Association, was angry but managed a polite rebuttal. Far less polite was Jack Chambers, one of the carriage operators. Petersen recalled that Chambers lectured him about "the commissioners sitting up on the hill in those cottages" who "knew or cared little about running a horse transportation company." Chambers asserted that despite decades of regulation, not only did the commissioners not know what they were talking about, the "commission had no right to tell an independent company how to do business." After the meeting had ended, Jack followed Petersen "and, with a grin, made a few concluding comments," one of which was that Woodfill could "kiss his ass."[49] On another occasion, Jack Gough, the city's mayor and a friend of the carriage operators, expressed his unhappiness with Woodfill in less graphic language, saying he was imposing "highfalutin Grand Hotel" standards on the carriage drivers with his insistence on "fancy uniforms."[50]

In truth, giving Woodfill all that he wanted raised the real possibility of putting the carriage tour company out of business. The Mackinac Island Carriage Tours had a difficult time finding and keeping drivers. By the 1950s, the number of people who knew how to handle a team of horses was relatively small. If the company cracked down too hard, the drivers might quit. Should that happen, Eugene Petersen admitted, "the company was dead in the box office."[51]

Woodfill was uninterested in the staffing problems of the carriage operators and proceeded to crack down particularly hard on what seems to have annoyed him the most: uniforms. The commission empowered Woodfill, as its chair, and also the Park Director Eugene Petersen, to immediately seize the operating license and ban from the park any carriage whose driver was not wearing the appropriate uniform or was wearing a uniform that was not "properly dry cleaned and pressed."[52] Insofar as the commission was concerned, carriage drivers would not only wear their uniform, they would wear it in a way satisfactory to Mr. Woodfill.[53]

While the Woodfill years were ones of extreme conflict, the extended history of battles between the carriage operators and the commission did not end when Woodfill was replaced by Walter J. Murray as the chair of the commission.[54] In 1967, the question of uniforms emerged again.[55] In 1974, the commission returned the time-worn topics of "flagrant violations of terms and conditions pertaining to solicitation of tips, wearing of proper uniforms, and encouraging passengers to visit Fort Mackinac."[56] The conflict between the commission and the carriage operators was never ending. Through it, it is possible to see the difficulty in finding common ground when decisions regarding fees and regulations to protect the nature of the tourist experience met the reality of making a living at a seasonal tourist trade.

Eugene Petersen expressed this problem more broadly when he wrote:

> Beneath the charm of Mackinac Island's historical and natural beauty was a political underground that resisted change or, worse, was indifferent to it. Vested interests with a large financial stake in serving the thousands of summer visitors welcomed our effort at Fort Mackinac, but only if it didn't affect their narrow interests.[57]

The vested and often narrow interests seen on the island also existed in St. Ignace and Mackinaw City. There were many efforts made to overcome these interests on the island, as well as in St. Ignace and Mackinaw City, but none was particularly successful. For example, in 1923, the Mackinac Island Association was formed as "a clearinghouse of ideas for the preservation and advancement of Mackinac Island." Everyone was welcome and the group was presented as a place where ideas could be discussed and resolved. Or, as a newspaper story had it, a place where "they can be dispensed with without hurting anybody's feelings."[58] On the mainland, the first cooperative effort to promote tourism in St. Ignace appears to have occurred in 1901, while an association to promote tourism existed in Mackinaw City by 1926.[59]

The problem with these efforts toward unity was that they rarely lasted very long. In 1926, the Mackinac Island Association embraced the park commission's controversial plan to restore the island's waterfront by removing the buildings between Main Street and the beach.[60] It was a controversy that was tearing apart the community, and taking a stand on it destroyed the organization. The window for thoughtful discussion and resolving issues without hurt feelings that had begun when the organization had formed lasted approximately three years.[61]

The same pattern occurred in St. Ignace and Mackinaw City. In St. Ignace, despite whatever fruits had been realized by the 1901 effort to organize in support of tourism, the tree had withered and died in a decade or less. A new organization was put together in 1916.[62] It too, however, bore little fruit. In May of 1922, St. Ignace created a "Booster Organization," with the goal of making every citizen a booster of tourism.[63] But a decade later, the boost had gone bust, and in 1933, St. Ignace was again trying to organize a body to coordinate tourism, this time as a Chamber of Commerce.[64] Like its predecessors, the Chamber of Commerce was not successful at creating a unified approach to tourism within the community. As a result, just after World War II ended, a new organization was created to try to accomplish this goal: the St. Ignace Junior Chamber of Commerce. In words that paraphrased what had been said in 1901, 1916, 1922, and 1933, the Junior Chamber of Commerce wrote,

> The cooperation of every individual must be sought in the creation of good will towards St. Ignace. As the gateway to the Upper Peninsula, we are directly responsible for that all-important first impression the tourist carries with him through the whole of northern Michigan. Our dealings with these pleasure-seeking people should be in SERVICE not in EXPLOITATION."[65]

In Mackinaw City, the organization that existed in 1926 to encourage tourism likely became the Chamber of Commerce that organized the 1933 commemoration of the 1763 battle at Fort Michilimackinac.[66] It eventually died an undocumented death. In 1955, the city's merchants formed a new Chamber of Commerce.[67] But if the Chamber's goal was to create a unified approach to tourism, it became quickly obvious that this was not to be accomplished. Of time spent at a late 1950s dinner that was to cement unity between the commission and the leaders of Mackinaw City, Eugene Petersen would recall that it was a pleasant evening. But, he wrote, after everyone went home, "neither Woodfill nor I had any illusions that relationships with the Mackinaw City politicians would be smooth."[68] Indeed, the relationship during the late 1950s and early 1960s was extremely rocky.[69]

Could it have been otherwise? In 1940, approximately four hundred people gathered at the Grand Hotel to discuss tourism.[70] Roger Andrews, the managing editor of the *Detroit Times* who served two terms on the park commission during the 1930s and was elected commission chair, attended the meeting and used it as an opportunity to promote unity: "We can work together and win. We can be selfish and lose."[71] Over the long haul, most of the tourist industry acted in a way Andrews would call selfish.

The Reluctance to Embrace Tourism as an Economic Mainstay

There are many ways to explain what Eugene Petersen saw as vested interests protecting their narrow economic concerns. One way is to agree with Roger Andrews and characterize the business community's behavior as selfish. Another viewpoint was expressed by Harry Ryba, who arrived on Mackinac Island in 1959. He did not see selfish people but rather people who lacked vision.[72] But neither Andrews or Ryba, or for that matter even Petersen, fully captured the issue.

Tourism in Michigan is seasonal. The summer season is full of visitors, but the winter months see few guests. The reality is that most people in the straits-area tourist industry must make enough money in six months to support themselves for a full year.[73] Not only is tourism in Michigan seasonal, it is also subject to substantial variability from year to year. When times are good, taking a week off to see the sites appeals to everyone. But in tough economic times, taking a vacation is among the first things a person or family can forgo. When the economy booms, tourism particularly benefits. When the economy declines, tourism is particularly hard hit.

Straits-area population statistics, which can be used to measure long-term trends in a local economy, reveal that even in the best of times, tourism did not create a vibrant regional economy. Over approximately a century, the local tourist industry was not capable of generating a large number of attractive jobs that could sustain a large number of workers that lived in the area.[74]

In 1884, a total of 3,184 people lived on Mackinac Island, in St. Ignace, and in Mackinaw City. In 1990 the three communities' total population was 3,914. This modest growth over approximately a century of what was, and remained, a small population hides important changes. Variability in population between 1884 and 1920 was caused by the rise and fall of the lumber industry in St. Ignace, as well as substantial changes in the tourist industry on Mackinac Island. From 1920 to 1960, the region's population grew decade by decade, reflecting a generally successful and, on the whole, expanding straits-area tourist business. Population peaked in

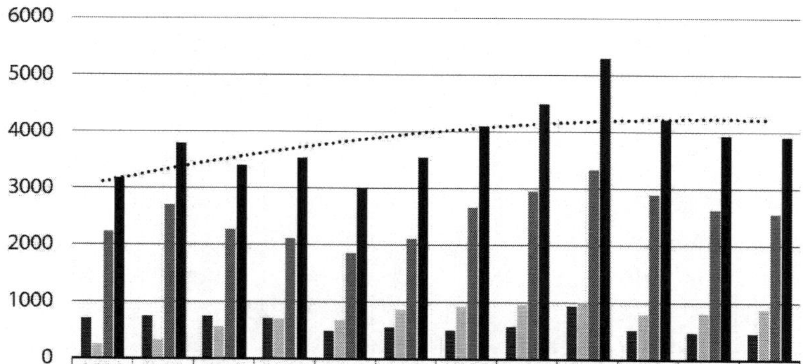

	1884	1890	1900	1910	1920	1930	1940	1950	1960	1970	1980	1990
■ Mack Island	703	750	564	714	493	566	508	572	942	517	479	469
■ Mack City	245	333	564	697	679	875	922	970	1023	810	820	884
■ St. Ignace	2236	2704	2271	2118	1852	2109	2669	2946	3334	2892	2632	2561
■ Total	3184	3787	3399	3529	3024	3550	4099	4488	5299	4219	3931	3914

Mack Island — Mack City — St. Ignace — Total ········ Poly. (Total)

Population Statistics for 1930, 1940, and 1950 were derived from US Department of Commerce, Bureau of the Census, *1950 Census of the Population*, vol. 1, *Number of Inhabitants: Michigan*, 22–15, 22–16, 22–18. Figures for 1960 are from the US Department of Commerce, Bureau of the Census, *1960 Census, Characteristics of the Population*, part 24, *Michigan*, table 7, 24–16, 24–18, https://www2.census.gov/library/publications/decennial/1960/population-volume-1/37722966v1p24ch2.pdf; figures for 1970 are from US Department of Commerce, Bureau of the Census, *1970 Census, Population*, vol. 1, *Characteristics of the Population,* part 24, *Michigan*, table 6, https://www.census.gov/library/publications/1973/dec/population-volume-1.html#par_list. Note the US Department of Commerce, Bureau of the Census, *Advance Report, Michigan, January 1971* (PC(v1) -24), table 2, pages 5–6, 10, has a different figure for the population of Mackinac Island in 1960 that does not appear in the census counts for 1960 and 1970. I have ignored this variant. The count for 1980 is from US Department of Commerce, Bureau of the Census, *Local Population Estimates, East North Central Series*, part 26, no. 84 ENC-SC, issued June 1986, table 1, pages 62, 64, 70. Figures for 1990 are from "Mackinaw City Population History, 1990–2023," Biggest US Cities, https://www.biggestuscities.com; "Mackinac Island Population History 1990–2023," Biggest US Cities, https://www.biggestuscities.com/city/mackinac-island-michigan; and "St. Ignace Population History, 1990–2023," Biggest US Cities, https://www.biggestuscities.com/city/st-ignace-michigan. For the purposes of this book, the small populations of nearby townships are not included.

1960 at 5,210, when optimism regarding regional tourism created by the opening of the Mackinac Bridge was at its height.[75]

That this optimism was false is demonstrated by the dramatic decline in the region's population between 1960 and 1970, when the population dropped approximately 19 percent. The Mackinac Bridge did not create an economic boom at the Straits of Mackinac. Quite the opposite; the bridge allowed tourists to drive right through. Although the decline in population was slower between 1970 and 1990, the ongoing loss of population documents that the straits region's economy was declining rather than expanding. Working in an industry that expanded slowly at best and often declined, those in the region's tourist industry chose to play their cards very close to the vest. They protected their interests, as Petersen correctly saw, but did so not out of selfishness or shortsightedness, but more likely out of a deep sense of caution, bordering on fear, regarding what next year might bring.

Understanding the difficulty in making a living through tourism also explains why many people in the straits region preferred year-round enterprises, even if creating such industries would have a negative effect on tourism. In the nineteenth century, islanders fished; tourism was an afterthought turned to only when the commercial fishing industry collapsed. At least some of the island's conflict over Moral Re-Armament (MRA) in the 1960s was about individuals preferring the year-round income they received from MRA, regardless of the consequences of the movement on island tourism.[76] Similarly, in St. Ignace and in Mackinaw City, efforts to find some source of income other than tourism should be understood as an effort to ensure community residents of a year-round income. A factory cleaning and freezing fish on the St. Ignace waterfront may be unattractive to tourists, but it would pay wages every week to its workers, unlike a waterfront gift shop that needed workers only in the summer.

Economic reality, however, was something locals argued over, but not something about which tourists were usually interested. Whatever the actual economic situation of straits residents, tourists came to experience something else. What tourists came to see, W. Stewart Woodfill recognized, was most often built on "illusions" and the "impressions."[77]

Remembering this, it is best to end by remembering the earliest impressions of the Stratis of Mackinac, penned by Harriet Martineau. Her words were written more than a century and a half ago, but they are words that might just as well have been written yesterday, and they explain today, as they did then, why people visit the Straits of Mackinac.

> We were in great delight at having seen Mackinac, at having the possession of its singular imagery for life; but this delight was at present dashed with the sorrow of leaving it. I could not have believed how deeply it is possible to regret a place, after so brief an acquaintance with it.[78]

Notes

CHAPTER 1. THE PIONEER TOURISTS, 1836–1861

1. James L. Clayton, "The American Fur Company: The Final Years" (PhD diss., Cornell University, 1964), 317–325; George Cantor, *Explore Michigan: Mackinac* (University of Michigan Press, 2005), 12; Brian Leigh Dunnigan, *A Picturesque Situation: Mackinac before Photography* (Wayne State University Press, 2008), 198, 91; Phil Porter, *Mackinac: An Island Famous in These Regions* (Mackinac State Historic Parks, 1988), 33.
2. Edwin O. Wood, ed., *Historic Mackinac: The Historical, Picturesque, and Legendary Features of Mackinac Country*, 2 vols. (MacMillan, 1918), 2:255.
3. Harriet Martineau, *Society in America*, quoted in Wood, *Historic Mackinac*, 2:255–256. The most complete compilation of the pre–Civil War island narratives was assembled by Wood. His work serves as the most convenient source through which these publications can be found.
4. Martineau, *Society in America*, 2:265–266.
5. Martineau, *Society in America*, 2:263–264.

6. Martineau, *Society in America*, 2:267, 2:264.
7. Genesis 9:28–29, New International Version (NIV).
8. Martineau, *Society in America*, 2:268. Dunnigan, *Picturesque Situation*, 91, attributes the first statement about no one ever dying on Mackinac Island to Dr. Francis LeBarron, who visited in 1802. More than two centuries later, the joke continued to be repeated. For example, see the *Saline Observer*, May 2, 1957, 6: "Did you know that most of the graves on Mackinac Island are of doctors? The climate of the isle is so healthy, they all starve to death for lack of patients."
9. Martineau, *Society in America*, 2:264, 2:266.
10. Martineau, *Society in America*, 2:268–269.
11. Kristen L. Hoganson, *Consumers' Imperium: The Global Production of American Domesticity, 1865–1920* (University of North Carolina Press, 2007), 166.
12. Dunnigan, *Picturesque Situation*, 197.
13. Clara Thomas, "Anna Jameson: Art Historian and Critic," *Woman's Art Journal* 1, no. 1 (1980), 20–22; Clara Thomas, "Murphy, Anna Brownell (Jameson)," *Dictionary of Canadian Biography* 8, http://www.biographi.ca/en/bio/murphy_anna_brownell_8E.html.
14. Thomas, "Murphy, Anna Brownell (Jameson)."
15. Thomas, "Murphy, Anna Brownell (Jameson)"; Jennifer Henderson, *Settler Feminism and Race Making in Canada* (University of Toronto Press, 2003), 45.
16. Henderson, *Settler Feminism*, 56.
17. Anna Jameson, *Winter Studies and Summer Rambles in Canada*, in Wood, *Historic Mackinac*, 2:279, 280.
18. For a chapter-long discussion of the relationship between Jameson and Jane Johnston, who was a member of an important mixed-blood trading family, see Melissa Croghan, *Great Women of Mackinac, 1800–1950* (Michigan State University Press, 2023).
19. Jameson, *Winter Studies*, 2:288–289.
20. Jameson, *Winter Studies*, 2:294–295.
21. Frederick Marryat, *Diary in America*, ed. Jules Zanger (Nicholas Vane, 1960), 9, 14.
22. Marryat, *Diary in America*, 122.
23. Marryat, *Diary in America*, 121. References to fairy isles were not entirely new. Anna Brownell Jameson referred to the islands in Lake Huron collectively as "fairy edens." Thomas, "Murphy, Anna Brownell (Jameson)."
24. Wood, *Historic Mackinac*, 2:361–362.
25. Margaret Fuller, *Summer on the Lakes*, in Wood, *Historic Mackinac*, 2:364, 2:368.

26. Fuller, *Summer on the Lakes*, 2:375.
27. Fuller *Summer on the Lakes*, 2:363.
28. Fuller, *Sumer on the Lakes*, 2:366–367.
29. Fuller, *Summer on the Lakes*, 2:369.
30. Fuller, *Summer on the Lakes*, 2:368–369.
31. Dunnigan, *Picturesque Situation*, ix, 170.
32. Dunnigan, *Picturesque Situation*, 248–249. Dunnigan also notes that the government work *Report on the Geology and Topography of a Portion of the Lake Superior Land District* by John Wells Foster and Josiah Dwight Whitney and published in 1851–1852 also included images of the island's geological features (265–267).
33. Dunnigan, *Picturesque Situation*, 294, 308–309.
34. Robert E. Benjamin, *Mackinac Island: Three Hundred Years of History* (Guelff, 1952), 43.
35. Dunnigan, *Picturesque Situation*, 145, 231.
36. Dunnigan, *Picturesque Situation*, 231; Willaim Henry Barrett, "The History of Transportation across the Straits of Mackinac" (MA thesis University of Hawaii, 1963), in *Memories of Mackinaw*, comp. Judy Ranville and Nancy Campbell (Little Traverse Printing, 1976), 121. There is some disagreement about how long it would take to load firewood onto a steamer, but it never took more than four hours.
37. Willis P. Hazard, *The American Guide-Book: Being a Handbook for Tourists and Travellers through Every Part of the United States* (D. Appleton, 1846), 140.
38. W. Williams, *Appletons' Southern and Western Travellers' Guide* (D. Appleton, 1850), 92–93.
39. W. Williams, *Appletons' New and Complete Guidebook for Travellers* (D. Appleton, 1851), 300.
40. Porter, *Mackinac*, 42.
41. Thomas Addison Richards, *Appletons' Illustrated Handbook of American Travel* (D. Appleton, 1857), 20, 43.
42. Dunnigan, *Picturesque Situation*, 198; Porter, *Mackinac*, 43.
43. Porter, *Mackinac*, 43; Dunnigan, *Picturesque Situation*, 231.
44. Porter, *Mackinac*, 43.
45. Fuller, *Summer on the Lakes*, 2:363.
46. William Cullen Bryant, *Letters of a Traveller*, in Wood, *Historic Mackinac*, 2:385.
47. Bryant, *Letters of a Traveller*, 2:396.
48. Louis Agassiz, *Lake Superior: Its Physical Character, Vegetation, and Animals Compared with Those of Other and Similar Regions*, with a narrative of the tour by J.

Elliot Cabot (Gould, Kendall and Lincoln, 1850), 22–23.

49. Daniel S. Curtiss, *Western Portraiture: An Emigrants' Guide*, in Wood, *Historic Mackinac*, 2:609.
50. Dunnigan reports that the community featured four bakeries in 1847. Dunnigan, *Picturesque Situation*, 231.
51. Dunnigan, *Picturesque Situation*, 245–246.
52. Porter, *Mackinac*, 42–43.
53. Charles Lanman, *A Summer in the Wilderness*, in Wood, *Historic Mackinac*, 2:601.
54. Jameson, *Winter Studies*, 2:287–288.
55. Bryant, *Letters of a Traveller*, 2:402.
56. John F. Sears, *Sacred Places: American Tourist Attractions in the Nineteenth Century* (Oxford University Press, 1989), 4, 6, 12. Sears notes that the grand tour of the day was codified by Timothy Dwight in 1821 and 1822. It consisted of taking in the scenery found in Connecticut and the Hudson River Valley, followed by a trip west through New York to Niagara Falls. Sears, *Sacred Places*, 49–50.

CHAPTER 2. BECOMING A FASHIONABLE RESORT, 1870–1894

1. John McCabe, *Grand Hotel Mackinac Island* (Lake Superior State College, 1987), 38; Cindy S. Aron, *Working at Play: A History of Vacations in the United States* (Oxford University Press, 1999), 53.
2. Phil Porter, *Mackinac: An Island Famous in These Regions* (Mackinac State Historic Parks, 1998), 37, 52; Melissa Croghan, *Great Women of Mackinac, 1800–1950* (Michigan State University Press, 2023), 157; *Centennial History of Cheboygan County* (repr. Beard, 1976; Northern Tribune Print, 1876, 69). The *Centennial History of Cheboygan* lists two island establishments that were still in the fishing business in 1876.
3. Aron, *Working at Play*, 15, 21.
4. Marion Morse Davis, "Stories of Saint Helena Island," *Michigan History Magazine* 10 (1926), 420–421, 439–441; Strickland, *Old Mackinaw*, in *Historic Mackinac: The Historical, Picturesque, and Legendary Features of Mackinac Country*, ed. Edwin O. Wood (MacMillan Company, 1918), 2:574.
5. David G. Schuster, "Neurasthenia and a Modernizing America," *JAMA 290, no. 17* (2003): 2327–2328; Gail Bederman, *Manliness and Civilization: A Cultural History of Gender and Race in the United States, 1880–1917* (University of Chicago Press, 1995),

84–87. Neurasthenia is no longer recognized as a diagnosis.

6. Mark Jackson, *Allergy: The History of a Modern Malady* (Reaktion Books, 2006), 57–64, 61.
7. John F. Sears, *Sacred Places: American Tourist Attractions in the Nineteenth Century* (Oxford University Press, 1989), 37–38, 176.
8. Aron, *Working at Play*, 45, 53.
9. *Appletons' Hand-Book of American Travel, Western Tour* (New York, 1871), 140–141.
10. William Cullen Bryant, ed., *Picturesque America or the Land We Live In*, 2 vols. (New York, 1872–1874), 1:279–291, quotation at 1:282. The first volume is important for singling out Mackinac Island among Great Lakes sites. The only other Great Lakes locations in volume 1 were Niagara Falls, the south shore of Lake Erie, and Lake Superior's Pictured Rocks, along with nearby Munising harbor. Volume 2 included only the cities of Chicago and Milwaukee from among Great Lakes locales. That Mackinac Island was so favored may be explained because the volume's editor, William Cullen Bryant, had visited Mackinac Island and published very favorable words about its natural beauty in 1846.
11. McCabe, *Grand Hotel Mackinac Island*, 14–15, 24, quotation at 15; Eugene Petersen, *Mackinac's Grand Hotel: The Early Years* (Historical Society of Michigan, 1988), 2–3; Steven C. Brisson, *Downtown Mackinac Island: An Album of Historic Photos* (Mackinac Island State Park Commission, 2015), 4.
12. John Disturnell, *Island of Mackinac, Giving a Description of All the Objects of Interest and Places of Resort* (Philadelphia, 1875) 24, 26–27. The *Centennial History of Cheboygan County* lists six hotels on the island and twelve businesses. Of the businesses, seven are selling general merchandise and seem to be catering to tourists. There is also a grocery/bakery store, two stores dealing with fish, and two stables. *Centennial History of Cheboygan County*, 69.
13. Disturnell, *Island of Mackinac*, 95–96. Additional information about stereo-optic view cards can be found in Porter, *Mackinac*, 52.
14. Steven C. Brisson, *Wish You Were Here: An Album of Vintage Mackinac Postcards* (Mackinac State Historic Parks, 2002), 7. M. Christine Byron and Thomas R. Wilson date the emergence of photography studies on the island to the 1880s. Byron and Wilson, *Vintage Views of the Mackinac Straits Region* (Arbutus Press, 2007), 104.
15. Porter, *Mackinac*, 53. Phil Porter, *Fudge: Mackinac's Sweet Souvenir* (Mackinac State Historic Parks, 2001), 8.
16. George Cantor, *Explore Michigan: Mackinac* (University of Michigan, 2005), 24.

17. Porter, *Fudge*, 9–10.
18. Porter, *Fudge*, 11–12. It should be noted that The Old Stand was the first store on the island to create candy before the visitors very eyes.
19. Andrew Watson, *Making Muskoka Tourism, Rural Identity, and Sustainability* (University of British Columbia Press, 2022), 80.
20. Aron, *Working at Play*, 3, 49.
21. Graydon M. Meints, *The Fishing Line: A History of the Grand Rapids & Indiana Railroad* (Michigan State University Press, 2018), 78.
22. Meints, *Fishing Line*, 77, 69.
23. Aron, *Working at Play*, 45.
24. Le Roy Barnett, *Railroads in Michigan: A Catalog of Company Publications, 1836–1980* (Northern Michigan University, 1986), 79–91.
25. Grand Rapids & Indiana Railroad, *A Guide to the Health, Pleasure and Fishing Resorts of Northern Michigan* (Dygert, Buff & Rich, 1879), 46–50.
26. Grand Rapids & Indiana Railroad, *A Guide to the Health*, 46–50. To read Dr. Drake see Daniel Drake, *The Northern Lakes: A Summer Resort for Invalids of the South* (NP, 1842; repr. Torch Press, 1964).
27. Railroad tourist publications were nationwide phenomena. One early example is *The Central Pacific Railroad: A Trip across the North American Continent from Ogden to San Francisco* (Central Pacific Railroad Company, 1871). On the other side of America, in 1875 the Pennsylvania Railroad published *The Pennsylvania Railroad: Its Origin, Construction, Condition, and Connections; Embracing Historical, Descriptive, and Statistical Notices of Cities, Towns, Villages, Stations, Industries, and Objects of Interest on Its Various Lines in Pennsylvania and New Jersey* (Pennsylvania Railroad, 1875). A later example, published in 1893, by the Northern Pacific Railroad Company was *Six Thousand Miles through Wonderland: Being a Description of the Marvelous Region Traversed by the Norther Pacific Railroad* (Rand, McNally, 1893).
28. Le Roy Barnett, *Shipping Literature of the Great Lakes* (Michigan State University Press, 1992), 140, 1.
29. *Ward's Lake Superior Line, Grand Pleasure Excursion to Lake Superior on the New and Superb Steamer Planet* (1855), in Barnett, *Shipping Literature*, 124.
30. Barnett, *Shipping Literature*, 83.
31. McCabe, *Grand Hotel*, 24.
32. Barnett, *Shipping Literature*, 57–64.
33. David D. Anderson, "John Disturnell Introduces the Great Lakes to America," *Inland Seas* 18 (Summer 1962): 106.

34. Dwight H. Kelton, *Annals of Fort Mackinac* (Chicago, 1882).
35. Kelton, *Annals*, 3.
36. Kelton, *Annals*. For information regarding Native legends, see pages 5–10. For information regarding the history of the island and the Great Lakes, see pages 30–43. A great deal of island history is also found under headings such as "History of the National Park" on pages 17–23 and "American Fur Co." on pages 51–55. Kelton also composed long lists of names including individuals married on the island (pages 58–69), Indian agents that served on the island (15), and Catholic priests that were stationed on the island (pages 56–57). The list of scenic attractions can be found on pages 98–111.
37. Worldcat lists editions published in 1886, 1887, 1888, 1892, 1893, and 1895. The Clarke Historical Library, Central Michigan University, holds, in addition to the volumes found in Worldcat, editions published in 1882, 1883, 1884, and 1894.
38. Watson, *Making Muskoka*, 79.
39. "A Brief History of the National Parks," Library of Congress, Washington, DC, https://www.loc.gov/collections/national-parks-maps/articles-and-essays/brief-history-of-the-national-parks/.
40. Thomas White Ferry, *Biographical Directory of the United States Congress*, https://bioguideretro.congress.gov/Home/MemberDetails?memIndex=f000095.
41. "Speech of the Honorable T.W. Ferry, United States Senator from Michigan, in the Senate, March 11, 1873," as printed in the *Congressional Record Containing the Proceedings and Debates of the 43rd Congress, Special Session of the Senate of the United States, March 4 to 26, 1873*, vol. 1 (US Congress, 1874), 39.
42. Sears, *Sacred Spaces*, 30.
43. Sears, *Sacred Spaces*, 184.
44. Phil Porter, *A Desirable Station: Soldier Life at Fort Mackinac, 1867–1895* (Mackinac State Historic Parks, 2003), 17.
45. Robert E. Benjamin, *Mackinac Island: Three Hundred Years of History* (Guelff, 1952), 44.
46. Keith R. Widder, "Mackinac National Park 1875–1975," *Reports in Mackinac History and Archaeology*, no. 4 (Mackinac State Park Commission, 1975), 14.
47. Widder, "Mackinac National Park," 4–6.
48. Sears, *Sacred Places*, 177.
49. Porter, *Desirable Station*, 17.
50. Widder, "Mackinac National Park," 17–20.
51. Phil Porter, *View from the Veranda: The History and Architecture of the Summer*

Cottages on Mackinac Island (Mackinac Island State Park Commission, 1981), 6.

52. Widder, "Mackinac National Park," 18–19. For a thorough description of the cottages' architectural styles as well as many photographs of the structures exteriors and interiors, see Porter, *View from the Veranda*, 38–62.
53. McCabe, *Grand Hotel*, 17.
54. "Cudahy Packing Co.," *Encyclopedia of Chicago*, http://www.encyclopedia.chicagohistory.org/pages/2635.html.
55. Porter, *Mackinac*, 63.
56. Mary Jane Doerr, *Bay View, an American Idea* (Priscilla, 2010), 19–23.
57. Meints, *Fishing Line*, 119.
58. Petersen, *Mackinac's Grand Hotel*, 2, 3.
59. Robert C. Ritchie, *The Lure of the Beach: A Global History* (University of California Press, 2021), 76–77.
60. Sears, *Sacred Places*, 66–68.
61. Petersen, *Mackinac's Grand Hotel*, 2–3.
62. McCabe, *Grand Hotel*, 21, 24, 33.
63. Benjamin, *Mackinac Island*, 56.
64. Petersen, *Mackinac's Grand Hotel*, 8–9.
65. Porter, *View from the Veranda*, 34.
66. Petersen, *Mackinac's Grand Hotel*, 5.
67. Brisson, *Downtown Mackinac Island*, 4.
68. Aron, *Working at Play*, 53.
69. Widder, "Mackinac National Park," 42. For a different view of the reasons why Mackinac Island became a national park than presented in this chapter, see Daniel McMahon, "The Role of Security Concerns in the of Mackinac National Park," *Michigan Historical Review* 50, no. 2 (Fall 2024), 87–109.

CHAPTER 3. MAKING THE PEOPLE'S RESORT, 1895–1930

1. Phil Porter, *A Desirable Station: Soldier Life at Fort Mackinac, 1867–1895* (Mackinac State Historic Parks, 2003), 23.
2. *Journal of the Senate of the State of Michigan, 1895*, 2 vols. (Lansing, 1895), 1:211; *Journal of the House of Representatives of the State of Michigan, 1895*, 3 vols. (Lansing, 1895), 1:342. The legislature was somewhat more subtle than the island's residents. Instead of asking for the military to remain, the legislature asked the

federal government to continue operating the national park, leaving it to those in Washington to determine how to do that without the army being stationed on the island.

3. John F. Sears, *Sacred Places: American Tourist Attractions in the Nineteenth Century* (Oxford University Press, 1989), 128. In 1906, California would give back to the federal government the land ceded to it in 1864, placing it in the Yosemite National Park, which was founded in 1890 and which largely surrounded the California state park. Sears, *Sacred Places*, 130.
4. Porter, *Desirable Station*, 23; David A. Armour, *100 Years at Mackinac: A Centennial History of the Mackinac Island State Park Commission, 1895–1995* (Mackinac Island State Historic Parks, 1995), 8.
5. Sears, *Sacred Places*, 128.
6. *Journal of the Senate of the State of Michigan, 1895*, 1:275; *Journal of the House of Representatives of the State of Michigan, 1895*, 1:196–197.
7. Sears, *Sacred Places*, 184.
8. *Journal of the House of Representatives of the State of Michigan, 1895*, 1:796, 1:851–852, 1:942, 2:1072–1073; *Journal of the Senate of the State of Michigan, 1895*, 1:642–643, 701; Armour, *100 Years at Mackinac*, 8.
9. Public Act 222, *Public Acts of the Legislature of the State of Michigan Passed at the Regular Session of 1895* (Lansing, 1895), 514–515; Armour, *100 Years at Mackinac*, 8.
10. *Journal of the Senate of the State of Michigan, 1895*, 2:1624–1625.
11. Mackinac Island State Park Commission (hereafter, MISPC) minutes, July 11, 1895, 2, Keith R. Widder Library, MISPC Office, Mackinaw City, MI. A complete set of the commission minutes are available at the Keith R. Widder Library, MISPC Office, Mackinaw City, MI. Most of the minutes consulted for this book are also available at the Michigan History Room, Mackinaw Area Public Library, Mackinaw City, MI. The set of minutes in each library is identical. The minutes are peculiar in that, while arranged chronologically, each typescript page is continuously paginated as if in a book. For example, the minutes of February 26, 1934, begin on page 237. The minutes of the commission's next meeting, held on July 29, 1934, begin on page 239. For convenience, I cite both date of meeting and page number of the minutes. Armour, *100 Years at Mackinac*, 8.
12. MISPC minutes, August 12, 1896, 6; January 8, 1897, 8; April 27, 1897, 9; Keith R. Widder Library, MISPC Office, Mackinaw City, MI. At these meetings, the commissioners discussed problems with rent collection.
13. MISPC minutes, November 2, 1897, 14; May 10, 1898, 17; Keith R. Widder Library,

MISPC Office, Mackinaw City, MI.

14. Armour, *100 Years at Mackinac*, 13, 30.
15. Armour, *100 Years at Mackinac*, 11, 13.
16. Armour, *100 Years at Mackinac*, 12; Porter, *Desirable Station*, 24–25. In the late 1880s, the number of soldiers stationed at Fort Mackinac peaked at one hundred and fifteen.
17. *Journal of the Senate of the State of Michigan, 1897*, 2 vols. (Lansing, 1897), 1:341, 1:352, 1:721, 1:821, 2:1116; *Journal of the House of Representatives of the State of Michigan, 1897*, 3 vols. (Lansing, 1897), 2:1334, 2:1416–1417. Mackinac Island is part of Mackinac County.
18. *Journal of the House of Representatives of the State of Michigan, 1897*, 3:2102–2104. Over the years, the commission would repeatedly flirt with the idea of expanding the number of cottage lots available to rent. For example, a plan for the development of the park drawn up by landscape architect Warren H. Manning in 1915 suggested laying out so many rental sites would have prevented public access to large sections of the park. The commission never acted on Manning's recommendation. See Steve Brisson, "Mackinac Island State Park, Mackinac Island, Michigan," in *Warren H. Manning, Landscape Architect and Environmental Planner*, ed. Robin Karson, Jane Roy Brown, and Sarah Allaback (University of Georgia, 2017), 183.
19. *Journal of the House of Representatives of the State of Michigan, 1897*, 3:2573–2574.
20. Armour, *100 Years at Mackinac*, 24.
21. Armour, *100 Years at Mackinac*, 14.
22. Sears, *Sacred Places*, 184–189.
23. "Biography of George T. Arnold," in *Historic Michigan: Land of the Great Lakes*, by George Fuller, vol. 3, *A History of the Upper Peninsula of Michigan* (National Historical Association, n.d.), 3:299–300. (For the sake of clarity it should be noted that Fuller's *Historic Michigan* series had several published volumes all identified on the title page as volume 3 but with unique volume titles and describing different geographic parts of the state. Cited here is the volume 3 with the title, *A History of the Upper Peninsula*.) "Biographies of Mackinac County Michigan: George T. Arnold," Genealogy Trails History Group, https://www.genealogytrails.com/mich/mackinac/bioarnold.html. No information can be found regarding William M. Clark.
24. *New York Times*, October 15, 1896, 1.
25. Clarence M. Burton and M. Agnes Burton, *History of Wayne County and the City of*

Detroit, Michigan, 3 vols. (S. J. Clarke, 1930), 3:204–208; Albert Nelson Marquis, ed., *The Book of Detroiters: A Biographical Dictionary of Leading Living Men of the City of Detroit* (A. N. Marquis, 1908), 427–428.

26. Levi L. Barbour, "Peter White as Man and as Citizen," *Michigan Alumnus*, 15 (1909): 407–412. For a fulsome account of White's accomplishments, see Ralph D. Williams, *The Honorable Peter White: A Biographical Sketch of the Lake Superior Iron Country* (Penton, 1907).
27. Eugene T. Petersen, *Mackinac's Grand Hotel: The Early Years* (Historical Society of Michigan, 1988), 5.
28. Petersen, *Mackinac's Grand Hotel*, 2–3.
29. Petersen, *Mackinac's Grand Hotel*, 12; John McCabe, *Grand Hotel Mackinac Island* (Lake Superior State College, 1987), 40, 46.
30. Petersen, *Mackinac's Grand Hotel*, 12.
31. McCabe, *Grand Hotel*, 50; Petersen, *Mackinac's Grand Hotel*, 11–13.
32. McCabe, *Grand Hotel*, 61.
33. Warren James Belasco, *Americans on the Road: From Autocamp to Motel, 1910–1945* (MIT Press, 1979), 59–61.
34. Angeline Teal, as quoted in Phil Porter, *View from the Veranda: The History and Architecture of the Summer Cottages of Mackinac Island* (Mackinac Island State Park Commission, 1981), 33–34.
35. Petersen, *Mackinac's Grand Hotel*, 15.
36. McCabe, *Grand Hotel*, 66. The denomination of the bill at the end of the flagpole seems to have varied over time. Porter states the successful contestant claimed a five-dollar prize. Porter, *View from the Veranda*, 37.
37. McCabe, *Grand Hotel*, 68.
38. Petersen, *Mackinac's Grand Hotel*, 16–17.
39. McCabe, *Grand Hotel*, 85.
40. McCabe, *Grand Hotel*, 72.
41. Petersen, *Mackinac's Grand Hotel*, 13–14.
42. McCabe, *Grand Hotel*, 73.
43. Phil Porter, *Mackinac: An Island Famous in These Regions* (Mackinac State Historic Parks, 1998), 61.
44. MISPC minutes, August 15, 1898, 26, Keith R. Widder Library, MISPC Office, Mackinaw City, MI.
45. MISPC minutes, October 27, 1899, 27, Keith R. Widder Library, MISPC Office, Mackinaw City, MI. It should be said that the commission did not always side with

those who preferred the public use of park land. In 1915 the commission received a report from Warren H. Manning, a distinguished landscape architect it had employed to develop the park. Manning suggested park land to the west of the fort be developed into an extensive public recreational area with space for baseball, football, basketball, and other outdoor activities, including an amphitheater for public performances. Shortly after receiving this recommendation the commission leased the land to the Grand Hotel for inclusion in their privately run golf course. Brisson, "Mackinac Island State Park, Mackinac Island, Michigan," 183.

46. Armour, *100 Years at Mackinac*, 30. Marquette was a French missionary priest who in 1668 founded Michigan's first European settlement, a Catholic mission at Sault Ste. Marie. In 1671, he helped found another mission at the Straits of Mackinac, located on land which would eventually become part of St. Ignace. Marquette died in 1675, and his remains were eventually interred at the St. Ignace mission.
47. Armour, *100 Years at Mackinac*, 23; Pamela A. Piljac and Thomas M. Piljac, *Mackinac Island: Historic Frontier, Vacation Resort, Timeless Wonderland* (Bryce-Waterton Publications, 1988), 131.
48. Porter, *Mackinac*, 68.
49. *St. Ignace News*, February 13, 1919, 1, reprinting an undated article from the *Detroit News*.
50. This chapter will document the conflict between the carriage operators and the commission in the 1920s but there is a long list of additional references to the conflict between the carriage operators and the commission. See Eugene T. Petersen, *Inside Mackinac* (Inside Mackinac, 1990), 122–126; the *Republican News and St. Ignace Enterprise*, September 22, 1938, 1; May 20, 1948, 2; October 27, 1949, 1; November 8, 1951, 10. Official documentation includes the MISPC minutes, June 5, 1901, 37; May 14, 1903; October 2, 1917, 104; May 23, 1922; August 19, 1924, 135; October 30, 1924, 137; August 17, 1925, 142; January 28, 1927, 163; September 7, 1928, 173; April 27, 1929, 176; July 5, 1930, 201; May 5, 1931, 209; July 10, 1936, 259; October 21, 1941, 302; December 3, 1946; June 1, 1950, 379; September 1, 1951, 389; December 27, 1951, 391; January 7, 1958 468; December 1, 1958; August 30, 1962, 30, 783–784; October 4, 1962, 796; November 3, 1972, 1356–1357; December 1, 1972, 1362; May 13, 1974, 146–164; October 4, 1974, 1489; November 14, 1974, 1495; May 9, 1975, 1524–1528; and the "Annual Superintendent's Report to the Mackinac Island State Park Commission," December 2, 1960, found in Superintendent's Records, box 1, MISPC Archives, Keith R. Widder Library, MISPC Office, Mackinaw City, MI.
51. Joseph F. Davey, "Memories of Mackinac," unpublished manuscript, 1982, 15.

52. Armour, *100 Years at Mackinac*, 16.
53. *Detroit Free Press* (hereafter, *DFP*), December 13, 1986, 27. Citations from the *Detroit Free Press* through 1990 are taken from ProQuest Historical Newspapers Detroit Free Press. Please note page citations found in this book are to the page number given by the ProQuest database. The database numbers each day's newspaper sequentially, rather than as they were originally ordered in the print edition, where each section of the paper assigned a letter and each section began with a new page 1. For example, what might have been originally numbered as section C, page 12, in the print edition, became page 37 in the ProQuest database. I have chosen to use the ProQuest page numeration on the assumption that most users will access the paper online and thus this makes subsequent access easier.
54. Armour, *100 Years at Mackinac*, 16.
55. McCabe, *Grand Hotel*, 77, 80, 94; Grand Rapids and Indiana Railway, *Annual Report for the Year 1907* (Grand Rapids and Indiana Railway Co., 1908), 21.
56. McCabe, *Grand Hotel*, 94–95. Grand Rapids and Indiana Railway Company, *Eleventh Annual Report for the Year Ending December 31, 1906* (Grand Rapids and Indiana Railway Co., 1907), 21.
57. McCabe, *Grand Hotel*, 95; *DFP*, June 17, 1911, 6.
58. McCabe, *Grand Hotel*, 97.
59. John William O'Malley, "The Story of the West Baden Springs Hotel" (MA thesis, Loyola University of Chicago, 1957), 79–88, quotation at 84.
60. Russell Magnaghi, *Prohibition in the Upper Peninsula* (American Palate, 2017), 102.
61. Larry D. Engelmann, "O, Whisky: The History of Prohibition in Michigan" (PhD diss., University of Michigan, 1971), 93, 104, 137, 276, 304–313.
62. McCabe, *Grand Hotel*, 111. Despite the general tolerance of both gambling and liquor sales, Mackinac Island did have some Prohibition advocates. Mrs. Nellie O'Leary and her daughter were known for their outspoken opposition to the sale of alcohol and their willingness to take matters into their own hands. On one occasion, the two women disrupted business at a local saloon, with the daughter hurling a hammer at a large mirror behind the bar, breaking it, while the mother, a hatchet in hand, stood guard to ensure her daughter's safety. Mrs. O'Leary also threatened to go after the liquor trade at the Grand Hotel. *St. Ignace Enterprise*, April 5, 1917, 1.
63. Robert C. Ritchie, *The Lure of the Beach: A Global History* (University of California Press, 2021), 78.
64. *DFP*, August 25, 1919, 1; August 27, 1919, 22.
65. *DFP*, August 27, 1919, 22. The state's attorney general eventually ordered the

Mackinac County prosecutor to file charges.

66. *DFP*, August 25, 1919, 1.
67. *DFP*, August 27, 1919, 22.
68. *Isabella County Enterprise*, October 1, 1920, 8. Found in the Clarke Historical Library, Central Michigan University, Digital Michigan Newspaper Database, https://digmichnews.cmich.edu/.
69. Magnaghi, *Prohibition in the Upper Peninsula*, 103, 104.
70. McCabe, *Grand Hotel*, 107–108; Magnaghi, *Prohibition in the Upper Peninsula*, 56, 103.
71. Armour, *100 Years at Mackinac*, 26, 25.
72. Marguerite S. Shaffer, *See America First: Tourism and National Identity, 1880–1940* (Smithsonian Institution Press, 2001), 124–125.
73. Armour, *100 Years at Mackinac*, 39–40.
74. Armour, *100 Years at Mackinac*, 40, 39.
75. Armour, *100 Years at Mackinac*, 41.
76. *St. Ignace Enterprise*, March 30, 1922, 8.
77. Armour, *100 Years at Mackinac*, 41; *DFP*, April 28, 1922, 11; *St. Ignace Enterprise*, March 30, 1922, 8.
78. *St. Ignace News*, January 23, 1919, 1; February 13, 1919, 1. The State Park Commission, which would operate all of the state park except the Mackinac Island State Park and the Michilimackinac State Park was created by the legislature in 1919. "Showcasing the DNR," Michigan Department of Natural Resources, https://content.govdelivery.com/accounts/MIDNR/bulletins/23168a5.
79. *St. Ignace Enterprise*, July 5, 1928, 8. The commission purchased the Early farm, sometimes known as the Dousman Farm.
80. *St. Ignace Enterprise*, December 27, 1923, 1; Armour, *100 Years at Mackinac*, 25–26.
81. *DFP*, November 7, 1920, 16; *Cheboygan Democrat*, November 12, 1920, 1; *St. Ignace News*, March 16, 1922, 1. The *Cheboygan Democrat* is found in the Clarke Historical Library, Central Michigan University, Digital Michigan Newspaper Database, https://digmichnews.cmich.edu/.
82. Armour, *100 Years at Mackinac*, 26.
83. Armour, *100 Years at Mackinac*, 26; *St. Ignace Enterprise*, March 16, 1922, 1; January 18, 1923, 8; December 27, 1923; Armour, *100 Years at Mackinac*, 26–28.
84. *St. Ignace Enterprise*, June 7, 1923, 8; People v. Doud, 223 Mich. 120 (1923). It is not clear if a new trial was ever held.
85. *St. Ignace Enterprise*, May 31, 1923, 8.
86. *St. Ignace Enterprise*, August 2, 1923, 8.

87. *St. Ignace Enterprise*, August 9, 1923, 8.
88. Mackinac Island Civic Association, *Mackinac Island* (Mackinac Island Civic Association, 1926), 3–4.
89. Mackinac Island Civic Association. *Mackinac Island*, 5–6.
90. *St. Ignace Enterprise*, May 27, 1926, 8.

CHAPTER 4. NEW LEADERS AND NEW DIRECTIONS, 1929–1941

1. Pamela A. Piljac and Thomas M. Piljac, *Mackinac Island: Historic Frontier, Vacation Resort, Timeless Wonderland* (Bryce-Waterton Publications, 1988), 143.
2. Otto Lang interview, *Town Crier*, December 1973, 34; Otto W. Lang, "Reminisces," 1975, 28, Keith R. Widder Library, Mackinac Island State Park Commission (hereafter, MISPC) Office, Mackinaw City, MI.
3. Marvin May, "Reminisces of Marvin May," 1999, 5, 13, Keith R. Widder Library, MISPC, Mackinaw City, MI.
4. Warren James Belasco, *Americans on the Road: From Autocamp to Motel, 1910–1945* (MIT Press, 1979), 143, 146.
5. Cindy S. Aron, *Working at Play: A History of Vacations in the United States* (Oxford University Press, 1999), 308n4.
6. *Town Crier*, August 11, 1990, 4.
7. Lang, "Reminisces," 65. Lang served as the hotel's receiver.
8. *St. Ignace Enterprise*, June 26, 1930, 8. The hotel's management, in a show of bravado, soon thereafter announced all was well and the hotel was doing capacity business. *St. Ignace Enterprise*, July 24, 1930, 8. Subsequent events proved this assertion untrue.
9. *St. Ignace Enterprise*, May 21, 1931, 8. In the same article reporting the reduction in rates, it was also written that the Grand's management was expecting a "big season." As with the brave statements made in 1930, this proved to be nonsense.
10. John McCabe, *Grand Hotel Mackinac Island* (Lake Superior State College, 1987), 113–114; *Clare Sentinel*, November 13, 1931, 12. The *Clare Sentinel* is found in the Clarke Historical Library, Central Michigan University, Digital Michigan Newspaper Database, https://digmichnews.cmich.edu/.
11. *Town Crier*, August 7, 1982, 23. Lang had been appointed receiver because the Arnold Ferry Line was among the hotel's largest local creditors.

12. Noella La Chance Schenck, *Winter Park's Old Alabama Hotel* (Anna Publishing, 1982), 18.
13. *Detroit Free Press* (hereafter, *DFP*), July 10, 1932, 2, 10; July 11, 1932, 10; July 12, 1932, 4, 8; July 14, 1932, 11, 12; July 15, 1932, 7; July 16, 1932, 9; July 17, 1932, 4, 9; July 19, 1932, 1; July 20, 1932, 14; July 21, 1932, 5; July 22, 1932, 7, 10; July 24, 1932, 9; July 26, 1932, 12; July 27, 1932, 2, 7; July 28, 1932, 4; July 29, 1932, 4; July 30, 1932, 16; July 31, 1932, 2, 5, 8, 9. The campaign included a series of positive articles. The articles invariably talked about how much those who had taken the trip enjoyed it, and they were often paired with pictures of the happy Island visitors and reminders to sign up soon for the next trip. Accompanying the articles were regularly printed, conventional display ads.
14. *St. Ignace Enterprise*, July 28, 1932, 8.
15. Schenck, *Winter Park's Old Alabama Hotel*, 18.
16. McCabe, *Grand Hotel*, 114.
17. H. Morgan Haskell, *W. Stewart Woodfill: Master of Mackinac's Grand Hotel* (H. Morgan Haskell, 2001), 59.
18. McCabe, *Grand Hotel*, xi.
19. Otto Lang, *Town Crier*, December 1973, 34.
20. McCabe, *Grand Hotel*, 113; Haskell, *Woodfill*, 58.
21. Henry G. Alsberg, ed., *The American Guide: A Source Book and Complete Travel Guide for the United States* (Hastings House, 1949), 542.
22. Haskell, *Woodfill*, 68.
23. Otto Lang, *Town Crier*, December 1973, 34. Lang served as the court appointed receiver for the Grand Hotel from 1931 until 1933. *Town Crier*, July 31, 1982, 23.
24. McCabe, *Grand Hotel*, 125.
25. McCabe, *Grand Hotel*, 125, 135–136; Haskell, *Woodfill*, 135. It should be noted that Woodfill was not alone in complaining about the electric company. In 1933, Mackinac Island's municipal government was also involved in a rate dispute with the company because of what it believed to be unjust rates. They too played hardball with the electric company by threatening to construct a municipally owned electric power plant. They too eventually managed to come to terms with the electric company. See *Republican News and St. Ignace Enterprise*, November 23, 1933, 1; December 7, 1933, 1; January 11, 1934, 8; January 18, 1934, 8.
26. Haskell, *Woodfill*, 60.
27. "An Interview with R.D. Musser II," Grand Hotel, https://www.grandhotel.com/press-media/interview-r-d-musser-ii/.

28. R. Daniel Musser Jr., "Summer 2012: An Oral History," conversation with Phil Porter, Mackinac State Historic Parks (2021), 15, Keith R. Widder Library, MISPC Office, Mackinaw City, MI. Haskell records that when the entire staff of bellmen went on strike during the 1930s, Woodfill fired them all and then had them put off the Island. He improvised a new staff overnight. Haskell, *Woodfill*, 66–67.
29. *Republican News and St. Ignace Enterprise*, October 3, 1935, 8.
30. Details of the three-way deal between Woodfill, the Ballards, and LaChance are found in the *Town Crier*, August 7, 1982, 23. Otto Lang recalled that the bankruptcy settlement included a side deal regarding gambling revenue between LaChance, the Ballards, and Woodfill, although he did not report the details of the arrangement. Lang, "Reminisces," 64.
31. Musser, "Oral History," 18. The Ballards' continued involvement in the Grand after it was purchased by Woodfill is difficult to document, but there are intriguing hints that the family maintained ties to the hotel. The St. Ignace newspapers reported Olive Ballard, the last family member to own the hotel and the woman who had placed it in receivership, spent the 1934 season at the Grand and was also at the hotel in 1935 and 1939. Perhaps Ms. Ballard simply liked the Grand, but her ongoing presence does lead to speculation. *Republican News and St. Ignace Enterprise*, July 19, 1934, 10; August 8, 1935, 8; September 14, 1939, 7.
32. *Republican News and St. Ignace Enterprise*, July 14, 1939.
33. *Town Crier*, August 7, 1982, 23.
34. W. Stewart Woodfill to Eugene Petersen, November 10, 1967, Petersen Woodfill Correspondence, box 2, MISPC Archives, Keith R. Widder Library, MISPC Office, Mackinaw City, MI; *Town Crier*, August 7, 1982, 23.
35. McCabe, *Grand Hotel*, 135–136.
36. *Town Crier*, August 7, 1982, 18.
37. Wilfird Doyle, as cited in *Town Crier*, July 13, 1985, 2.
38. Porter, *Mackinac*, 74–75. MISPC minutes, June 21, 1935, 247 Keith R. Widder Library, MISPC Office, Mackinaw City, MI, documents a 25 percent reduction across the board in cottage rental fees. In 1936, the commission reduced the discount to 10 percent. MISPC minutes, February 27, 1936, Keith R. Widder Library, MISPC Office, Mackinaw City, MI, 255. A 10 percent discount for "prompt payment" was offered in 1937. MISPC minutes, June 12 1937, 266, Keith R. Widder Library, MISPC Office, Mackinaw City, MI.
39. Wilfird Doyle, *Town Crier*, July 13, 1985, 2.
40. *Town Crier*, June 4, 1988, 25. Laws were passed to permit these sales, a belated

authorization of actions that the commission had undertaken but about which questions had been raised.

41. Public Act 222, *Public Acts of the Legislature of the State of Michigan Passed at the Regular Session of 1895* (Lansing, 1895), 514.
42. Public Act 355, *Public Acts of the Legislature of the State of Michigan Passed at the Regular Session of 1927* (Robert Smith, 1927), 853.
43. The law came into effect June 2, 1927. *Public Act*, 355; *DFP*, July 25, 1927, 1. Much of the excitement surrounding the event dissipated when all three men announced, on July 24, that they would not be in attendance.
44. Information regarding the members of the Mackinac Island State Park Commission and their length of service was retrieved from what is commonly called the *Michigan Manua*. It is a volume published by the state every two years to coincide with each legislative term. Twenty-two consecutive volumes were consulted, beginning with the volume published for the years 1895–1896 and ending in the volume published for 1939–1940.
45. David A. Armour, *100 Years at Mackinac: A Centennial History of the Mackinac Island State Park Commission, 1895–1995* (Mackinac Island State Historic Parks, 1995).
46. Public Act 219, *Public and Local Acts of the Legislature of the State of Michigan Passed at the Regular and Extra Sessions of 1937* (Franklin DeKleine, 1937), 349–350.
47. The resignation of the commissioners appointed by Governor Murphy and the names of the individuals who replaced them is documented in the *Journal of the Senate of the State of Michigan, 1939 Regular Session*, 2 vols. (Franklin DeKleine Company, 1939), 1:47, 1:264, 1: 279, 1:289, 1:631, 1:632, 1:665. To further complicate matters, Governor Fitzgerald, who succeeded Murphy, died in office March 16, 1939, before filling all the vacancies on the park commission. Thus, some of those who replaced Governor Murphy's commissioners were appointed by Fitzgerald, others by his successor, Lt. Governor Luren Dickinson.
48. Woodfill was known to glare menacingly at, and often verbally reprimand, anyone who dared call it a "cane."
49. Haskell, *Woodfill*, 1.
50. Lewis A. Harding, ed., *History of Decatur County, Indiana: Its People, Industries, and Institutions* (B. F. Bowen, 1915), 622–624.
51. Haskell, *Woodfill*, 8–9.
52. McCabe, *Grand Hotel*, 109.
53. *St. Ignace Enterprise*, August 25, 1927, 1. LaChance would also run the hotel during the 1932 season.

54. Haskell, *Woodfill*, 46.
55. Haskel, *Woodfill*, 50. Why Woodfill, who carefully polished his public image, chose to express his admiration for Edward Ballard so strongly is mysterious, although it may have been that two contradictory urges were at work: his sense of proprietary versus his love of presenting himself as a buccaneer with the nerve to do what it took to prosper.
56. *St. Ignace Enterprise*, May 31, 1923, 8.
57. Haskell, *Woodfill*, 53, 153. It is possible that Woodfill was also interested in either obtaining a share in or purchasing the Ballard's French Lick hotel. Haskell, *Woodfill*, 154.
58. McCabe, *Grand Hotel*, 112–113.
59. *St. Ignace Enterprise*, June 11, 1931, 5; October 8, 1931, 8; October 27, 1932, 8. [*St. Ignace*] *Republican News*, April 30, 1932, 8.
60. Haskell, *Woodfill*, 58, 129; *St. Ignace Enterprise*, February 23, 1933, 1. Although the bid was submitted in February, court approval of the sale was not finalized until May. *St. Ignace Enterprise*, May 4, 1933, 4. Since the hotel technically was still in the hands of the receiver, Otto Lang, Woodfill leased the hotel from him for the 1933 season while the paperwork wound its way through the court, *St. Ignace Enterprise*, February 23, 1933, 1.
61. McCabe, *Grand Hotel*, 142.
62. By way of example, the identical story, including the quote cited above, ran in the *Oxford Leader*, July 12, 1935, on page 1 and the *Clare Sentinel*, July 12, 1935, page 1. A somewhat tardy *Saline Observer* did not run the identical piece until July 18, 1935, and, disappointingly to Woodfill, it ran on on page 4. All of the newspapers mentioned are found in the Clarke Historical Library, Central Michigan University, Digital Michigan Newspaper Database.
63. Petersen, *Inside Mackinac*, 117. Gloria Swanson starred in her first feature movie in 1919. By 1925, she was one of Hollywood's most bankable stars. She also received critical acclaim. She was nominated for an Academy Award as best actress in 1928 and in 1929.
64. McCabe, *Grand Hotel*, 133, 142, 148.
65. *Town Crier*, August 11, 1972, 8; *South Florida Sun Sentinel*, January 18, 1988, sec. Obituary; *Oxford Leader*, October 5, 1951, 5. The *Oxford Leader* is found in the Clarke Historical Library, Central Michigan University, Digital Michigan Newspaper Database.
66. *Town Crier*, June 4, 1988, 25.

67. *DFP*, May 21, 1961, 17.
68. *Town Crier*, September 1, 1972, 5.
69. MISPC minutes, December 4, 1934, 240, Keith R. Widder Library, MISPC Office, Mackinaw City, MI.
70. *Republican News and St. Ignace Enterprise*, November 22, 1934, 8.
71. *Town Crier*, August 12, 1962, 12.
72. *Republican News and St. Ignace Enterprise*, June 25, 1936, 6, notes the arrival of the Doyle's for the summer at the Schultz cottage, which they are apparently renting.
73. See for example, *Republican News and St. Ignace Enterprise*, March 5, 1936, 8.
74. *South Florida Sun Sentinel*, January 18, 1988, sec. Obituary.
75. Lang, "Reminisces," 13. The Arnold estate was complicated in that multiple properties were involved, some solely in Arnold's name, some held jointly by himself and his wife, and others solely in Mrs. Arnold's name.
76. Lang, "Reminisces," 14, 81; *Town Crier*, August 26, 1978, 6; December 16, 1978, 2. The acquisition of the line and other property by Lang and Brown involved settling with approximately twenty heirs and a ten-year buy out agreement. As had happened when Mr. Arnold died, settling the estate was complicated.
77. *Town Crier*, December 16, 1978, 2.
78. MISPC minutes, September 27, 1935, 249, 250, Keith R. Widder Library, MISPC Office, Mackinaw City, MI.
79. MISPC minutes, September 27, 1935, 251; October 29, 1935, 252; Keith R. Widder Library, MISPC Office, Mackinaw City, MI; *Town Crier*, June 21, 1986, 5.
80. MISPC minutes, January 15, 1936, 253, Keith R. Widder Library, MISPC Office, Mackinaw City, MI.
81. The US Army Corps of Engineers recorded the number of passengers visiting Mackinac Island from 1915 until 1989, with the exception of the years 1952 through 1956. The corps also recorded the number of passengers carried by the state auto ferries across the straits from 1927 until 1951. These statistics can be found in *The Report of the Chief of Engineers, United States Army, Commercial Statistics*, published annually through the year 1952. Thereafter the corps removed the commercial statistics from the *Chief Engineers Report* and placed them in a new annual publication, *Waterborne Commerce of the United States*, which first appeared in 1956, but did not begin including passenger traffic to Mackinac Island until 1957. The statistics found in this chart are drawn from these publications.
82. *DFP*, August 10, 1935, 6; *Republican News and St. Ignace Enterprise*, November 23, 1939, 8.

83. *Saline Observer*, August 31, 1939, 4, Clarke Historical Library, Central Michigan University, Digital Michigan Newspaper Database, https://digmichnews.cmich.edu/.
84. For contemporary printings of Alleman's article see, for example the *Saline Observer*, August 31, 1939, 4, and the *Suttons Bay Courier*, August 31, 1939. The Island would continue to be seen as expensive for decades to come. For example, the *Detroit Free Press* wrote in 1988, "There's only one rule you must obey when you come to Mackinac Island: Come with lots of money; leave with no money." *DFP*, September 25, 1988, 123. The *Suttons Bay Courier* is found in the Clarke Historical Library, Central Michigan University, Digital Michigan Newspaper Database.
85. MISPC minutes September 27, 1935, 250.
86. *DFP*, August 10, 1935, 6.
87. *DFP*, August 19, 1939, 4. The letter is signed "J.T."
88. *DFP*, July 21, 1940, 7.
89. Clarence C. Eby, *Mackinac County of the Straits Country: An Official Tourist Guide into the Upper Peninsula of Michigan* (Clarence C. Eby, 1926), 28.
90. John A. Jakle, *The Tourist: Travel in Twentieth-Century North America* (University of Nebraska Press, 1985), 152–153.
91. *Republican News and St. Ignace Enterprise*, November 7, 1935, 1.
92. Armour, *100 Years at Mackinac*, 46.
93. Report of a Joint Meeting of the Mackinac Island Bureau, September 1, 1929, found in folder "Misc," MISPC Superintendent's Records, box 1, MISPC Archives, Keith R. Widder Library, MISPC Office, Mackinaw City, MI.
94. Armour, *100 Years at Mackinac*, 49.
95. Superintendent's report, December 30, 1930, in Mackinac Island State Park Superintendent Records, box 1, MISPC Archives, Keith R. Widder Library, MISPC Office, Mackinaw City, MI.
96. Superintendent's report, November 30, 1928, and superintendent's report, January 1, 1930, in Mackinac Island State Park Superintendent Records, box 1, MISPC Archives, Keith R. Widder Library, MISPC Office, Mackinaw City, MI. Some sources, such as Armour, *100 Years at Mackinac*, 50, say the state auto ferries ran to Mackinac Island in 1928 but did not resume in 1929. Because the Armour volume lacks footnotes, I cannot examine the source material Armour used for his statement. I have chosen to rely upon the superintendent's reports regarding the years of service.
97. *St. Ignace Enterprise*, January 2, 1930, 1.

98. *St. Ignace Enterprise*, December 26, 1929, 8; January 2, 1930, 1.
99. *St. Ignace Enterprise*, December 26, 1929, 8. Insofar as I can determine, no referendum was every held.
100. *Clare Sentinel*, August 16, 1935, 12, Clarke Historical Library, Central Michigan University, Digital Michigan Newspaper Database, https://digmichnews.cmich.edu/.
101. *DFP*, August 10, 1935, 6.
102. *Saline Observer*, August 31, 1939, 4.
103. *Saline Observer*, August 31, 1939, 4.
104. *The Report of the Chief of Engineers, United States Army, Commercial Statistics 1931*, 842; and *Commercial Statistics, 1944*, 890. It was not much better in 1939 than in 1938. A total of 139,813 visitors came to the Island in 1938, growing to 198,245 in 1939. In comparison, the island had drawn more than 300,000 visitors a year between 1927 and 1929.
105. *St. Ignace Newspaper*, November 23, 1939, 8.
106. Lang, "Reminisces," 76.
107. *St. Ignace Enterprise*, January 2, 1930, 1.
108. Hendricks to Governor Brucker, May 5, 1931, in folder "Misc," Mackinac Island State Park Superintendent Records, box 1, MISPC Archives, Keith R. Widder Library, MISPC Office, Mackinaw City, MI.
109. MISPC minutes, February 26, 1934, 237, Keith R. Widder Library, MISPC Office, Mackinaw City, MI.
110. MISPC minutes, June 21, 1935, 247.
111. MISPC, "1937–1938 Report of the Mackinac Island State Park Commission of Michigan, 1938," 18.
112. MISPC, "1933 and 1934 Report of the Mackinac Island State Park Commission," report submitted December 1, 1934, 5–6.
113. Time would support Andrews point regarding the remoteness of Isle Royale. In 2017 the National Park Service reported 26,306 visitors to Isle Royale National Park, while the same year the *Detroit Free Press* estimated Mackinac Island attracted about one million visitors.
114. MISPC, "1933 and 1934 Report," 6–8.
115. *DFP*, July 1, 1934, 8.
116. McCabe, *Grand Hotel*, 130; MISPC minutes, February 26, 1934, 237.
117. *Town Crier*, August 12, 1955, 6.
118. *Town Crier*, December 16, 1978, 2.

119. "History of Colonial Williamsburg," Colonial Williamsburg, https://www.colonialwilliamsburg.org/learn/about-colonial-williamsburg/history/.
120. *DFP*, January 27, 1942, 11; February 1, 1941, 45; March 19, 1941, 1; Otto Lang, *Town Crier*, August 13, 1977, 23; Lang, "Reminisces," 96. The *Detroit Free Press* attributes the tax idea to Doyle, naming it after him in the story they print; however, Otto Lang says he was discussing the tax with Woodfill. In all likelihood, the idea developed as part of an ongoing conversation between all of them, with no way to discern who first suggested it.
121. *Republican News and St. Ignace Enterprise*, March 6, 1941, 1; *DFP*, July 13, 1943, 43.
122. Lang, "Reminisces," 35.
123. *Republican News and St. Ignace Enterprise*, March 20, 1941, 1.
124. Armour, *100 Years at Mackinac*, 64.
125. *DFP*, June 6, 1941, 7; June 7, 1941, 16; June 15, 1941, 55.
126. *DFP*, June 15, 1942, 55.
127. *DFP*, July 9, 1941, 10.
128. *Mackinac Island News*, July 26, 1941, 8.
129. *DFP*, July 7, 1941, 4, 16.
130. Gerald A. Ramin, "Tourism as a Major Resource of Mackinac Island" (Independent study, MS degree, Mankato State University, 1963), 43.
131. *DFP*, July 26, 1941, 1, 18.
132. *DFP*, June 22, 1941, 6; July 20, 1941, 44; July 24, 1941, 16.
133. Robert C. Ritchie, *The Lure of the Beach: A Global History* (University of California Press, 2021), 204.

CHAPTER 5. THE LEADERSHIP DIVIDED, 1941–1947

1. *Republican News and St. Ignace Enterprise*, March 13, 1941, 8.
2. A total of 285,633 visitors came to the island, the best since 1929. *Report of the Chief of Engineers, U.S. Army, 1935, Part 2 Commercial Statistics* (Government Printing Office, 1931), 842; *Annual Report of the Chief of Engineers, 1943, Part 2 Commercial Statistics* (Government Printing Office, 1944), 890; *Annual Report of the Chief of Engineers, 1953, Part 2 Commercial Statistics* (n.p., n.d.), 1212.
3. *Detroit Free Press* (hereafter, *DFP*), July 12, 1942, 6.
4. *DFP*, June 27, 1943, 46; McCabe, *Grand Hotel*, 154.
5. *DFP*, August 25, 1943, 3. The claim was untrue. In 1943, 100,866 people visited the

island. The year before, 160,830 had come. Both years had far less the 300,000 or more visitors who had come in the years just before the Depression.

6. In 1924, 82,518 visitors had taken a ferry to the island. Every year thereafter would be above 100,886 visitors.
7. *DFP*, August 1, 1943, 45.
8. *DFP*, May 9, 1943, 12.
9. "United States Home Front During World War II," Wikipedia, https://en.wikipedia.org/wiki/United_States_home_front_during_World_War_II.
10. *Courier-Northerner*, November 12, 1943, 6. The *Courier-Northerner* is found in the Clarke Historical Library, Central Michigan University, Digital Michigan Newspaper Database, https://digmichnews.cmich.edu/.
11. *DFP*, May 5, 1943, 17.
12. *DFP*, January 10, 1943, 7.
13. John McCabe, *Grand Hotel Mackinac Island* (Lake Superior State College, 1987), 154.
14. Marvin May, "Reminisces of Marvin May," 1999, 19, Keith R. Widder Library, Mackinac Island State Park Commission (hereafter, MISPC) Office, Mackinaw City, MI.
15. Cindy S. Aron, *Working at Play: A History of Vacations in the United States* (Oxford University Press, 1999), 238, 247. The advantages and disadvantages of paid vacations for industrial workers had been widely discussed during the 1920s, but in 1927, only about 5 percent of America's factory workers received a paid vacation. This changed radically in the late 1930s, with paid vacations often given by manufacturers in the hopes of keeping workers from unionizing. See Aron, *Working at Play*, 248–249.
16. Harvey H. Jackson III, *Redneck Riveria: An Insider's History of the Florida-Alabama Coast* (University of Georgia Press, 2012), 40.
17. *L'Anse Sentinel*, June 25, 1947. The *L'Anse Sentinel* is found in the Clarke Historical Library, Central Michigan University, Digital Michigan Newspaper Database, https://digmichnews.cmich.edu/.
18. *Annual Report of the Chief of Engineers, 1952, Part 2 Commercial Statistics* (Government Printing Office, 1953), 1212.
19. *DFP*, September 19, 1948, 20.
20. Otto Lang, "Reminisces," 1975, 79, Keith R. Widder Library, MISPC Office, Mackinaw City, MI.
21. *DFP*, February 1, 1942, 45.
22. *DFP*, April 1, 1954, 33.

23. Lang, "Reminisces," 79.
24. Lang, "Reminisces," 96; *DFP*, February 1, 1942, 45.
25. Lang, "Reminisces," 81.
26. *Republican News and St. Ignace Enterprise*, July 17, 1941, 8; September 2, 1941, 8. The bonds legality was always an issue. An act of the legislature, passed after the island voters had approved the law, was necessary to make clear the bonds were legal under Michigan law. See *Republican News and St. Ignace Enterprise*, May 15, 1941, 8; *DFP*, February 1, 1942, 45; *Republican News and St. Ignace Enterprise*, January 9, 1942, 8.
27. Lang, "Reminisces," 109. As it turned out, the bonds had been bought in their entirety by the Mott Foundation, a Michigan not-for-profit corporation. Apparently, the crisis that the default would have normally caused was averted when the foundation agreed to postpone payments and gave the Park and Harbor Commission the time needed for Lang to work out a way to obtain the money.
28. Lang, "Reminisces," 98–100.
29. *South Florida Sun Sentinel*, January 18, 1988, sec. Obituary.
30. *DFP*, October 28, 1950, 18.
31. *DFP*, December 13, 1942, 10.
32. David A. Armour, *100 Years at Mackinac: A Centennial History of the Mackinac Island State Park Commission, 1895–1995* (Mackinac Island State Historic Parks, 1995), 63. Public Act 30, *Public and Local Acts of the Legislature of the State of Michigan Passed at the Regular Session of 1941* (Franklin DeKleine, 1941), 28.
33. Armour, *100 Years at Mackinac*, 63, 64; *DFP*, December 13, 1942, 10.
34. *DFP*, December 13, 1942, 10.
35. Armour, *100 Years at Mackinac*, 65.
36. Armour, *100 Years at Mackinac*, 65.
37. *Saline Observer*, July 12, 1945, 6.
38. Armour, *100 Years at Mackinac*, 65–66. Carl D. Nold, "The Michigan Governor's Summer Residence," *Mackinac History* 5, leaflet no. 6 (2024), 8. Many years later Doyle would pat himself on the back for engineering the purchase of the Governor's Residence on the island. *Mackinac Island Town Crier*, July 6, 1985, 3.
39. Governor Williams purchased a cottage on the island that was on park land shortly before his term of office ended. Governor William G. Milliken (term of office 1969–1983) had a cottage built on the island on privately owned land he purchased.
40. MISPC minutes, May 9, 1975, 1520.

41. H. Morgan. Haskell, *W. Stewart Woodfill Woodfill: Master of Mackinac's Grand Hotel* (H. Morgan Haskell, 2001), 76–78.
42. By way of example of the almost daily references to the Mackinac Conference in the press, in September 1944 the gathering s mentioned in the *DFP* on September 1, pages 8, 12; September 3, page 1; September 4, pages 1, 11, 15; September 5, pages 1–2, 41; September 6, pages 1, 4, 5, 19; September 7, page 1; September 8, pages 1, 5, 15, 19; September 9, page 16; September 10, pages 5, 17; September 11, pages 6, 15: September 12, pages 3, 6; September 14, page 16; September 16, page 18; September 18, page 1; September 19, page 45; September 20, page 5; September 23, page 15; September 26, page 45; September 29, page 6.
43. *DFP*, June 30, 1945, 1.
44. *DFP*, May 23, 1945, 2.
45. *DFP*, June 12, 1945, 1.
46. *DFP*, July 1, 1945, 3.
47. *Saline Observer*, July 12, 1945, 6.
48. *DFP*, May 11, 1946, 11.
49. *DFP*, July 22, 1946, 13.
50. *DFP*, July 22, 1946, 13.
51. Charles Deitz, MGM Pictures to Gov. Harry Kelly, July 14, 1946, in the Petersen Woodfill Correspondence, box 1, MISPC Archives, Keith R. Widder Library, MISPC Office, Mackinaw City, MI.
52. R. Daniel Musser Jr., "Summer 2012: An Oral History," conversation conducted by Phil Porter, director, Mackinac State Historic Parks (2021), 29, Keith R. Widder Library, MISPC Office, Mackinaw City, MI.
53. W. Stewart Woodfill to William Doyle, July 14, 1946, Petersen Woodfill Correspondence, box 1, MISPC Archives, Keith R. Widder Library, MISPC Office, Mackinaw City, MI.
54. W. Stewart Woodfill to William Doyle, July 16, 1946, Petersen Woodfill Correspondence, box 1, MISPC Archives, Keith R. Widder Library, MISPC Office, Mackinaw City, MI.
55. Musser, "Summer 2012," 29.
56. *DFP*, August 24, 1946, 4.
57. Donald Fry Prather, *There Will Always Be a Mackinac Race* (D. F. Keller, 1925), 11.
58. *DFP*, July 11, 1946, 15.
59. *DFP*, July 16, 1946, 12; McCabe, *Grand Hotel*, 160.
60. Prather, *There Will Always Be a Mackinac Race*, 59–60. All quotations found on

these pages.

61. McCabe, *Grand Hotel*, 160; *DFP*, August 7, 1946, 4, 24.
62. *DFP*, August 23, 1946, 3; August 24, 1946, 4. The park commission, of which Doyle was a member, also offered to help restore harmony. MISPC minutes, December 3, 1946, 337, Keith R. Widder Library, MISPC Office, Mackinaw City, MI.
63. McCabe, *Grand Hotel*, 162, 163. The yacht race business would remain a lucrative event for many years. In 1986, for example, the *Detroit Free Press* reported it was not unusual for a captain to spend thousands of dollars on breakfast at the Chippewa Hotel, a favorite spot for race participants. Breakfast at the hotel in 1986 cost five dollars. Most of the remaining bill seems to have come from the many bottles of champagne that were consumed. *DFP*, June 29, 1986, 82.
64. McCabe, *Grand Hotel*, 163–164.
65. All of Doyle's quotes appear in *Republican News and St. Ignace Enterprise*, April 3, 1947, 8.
66. *Republican News and St. Ignace Enterprise*, April 24, 1947, 8; *DFP*, April 22, 1947. 3.
67. Musser, "Summer 2012," 29.
68. *Republican News and St. Ignace Enterprise*, March 11, 1948, 1.
69. *Republican News and St. Ignace Enterprise*, April 3, 1947, 8.
70. James P. Dunnigan, "Around My World in Eighty Years," unpublished memoir (1990), 30, Keith R. Widder Library, MISPC Office, Mackinaw City, MI.
71. *DFP*, March 5, 1948, 11; September 30, 1948, 3.
72. *Town Crier* June 1, 1984, 19.
73. *Commercial Statistics: Water-Borne Commerce of the United States for the Calendar Year 1944* (Government Printing Office, 1946), 974, 977; *Commercial Statistics Water-Borne Commerce of the United States for the Calendar Year 1952* (US Printing Office, 1954), 1188, 1191.
74. *DFP*, December 18, 1942, 8; Armour, *100 Years at Mackinac*, 112.

CHAPTER 6. THE GOVERNOR CHARTS A NEW COURSE, 1948–1961

1. Stephen Franklin, *Chicago Tribune*, February 3, 1988.
2. Thomas Noer, *Soapy: A Biography of G. Mennen Williams* (University of Michigan Press, 2005), 77–78.
3. Insofar as can be determined, the Republicans never seriously considered

appropriating the funds required to build the Mackinac Bridge for highway projects in the Detroit area.

4. *Detroit Free Press* (hereafter, *DFP*), February 18, 1950, 6. For an opinion piece on the subject that also supports the idea of using the money for the proposed bridge to improve highways in the Detroit area, see *DFP*, January 9, 1950, 6.
5. Statement of Mr. Ellsworth, read into the *Journal of the Senate of the State of Michigan, 1950 Extra Session* (Franklin DeKleine, 1950), 163–164.
6. The 1950 population numbers for the Upper and Lower Peninsula were calculated by combining individual county population totals for 1950, which were printed in the *Michigan Official Directory and Legislative Manual, 1951* (State of Michigan, 1951), 412–415.
7. *DFP*, February 16, 1949, 7. Bids for the new ferry were accepted by the state in September 1949. See *DFP*, September 4, 1949, 3.
8. *Republican News and St. Ignace Enterprise*, January 17, 1952, 1.
9. Noer, *Soapy*, 120.
10. Lawrence A. Rubin, *Bridging the Straits: The Story of Mighty Mac* (Wayne State University Press, 1985), 7.
11. H. Morgan Haskell, *W. Stewart Woodfill: Master of Mackinac's Grand Hotel* (H. Morgan Haskell, 2001), 87, 86.
12. Haskell, *Woodfill*, 88–89.
13. Public Act 35, *Public Acts of the Legislature of the State of Michigan Passed at the Extra Sessions of 1933–1934* (Franklin DeKleine, 1934), 119; Public Act 300, *Public and Local Acts of the Legislature of the State of Michigan Passed at the Regular Session of 1947* (Franklin DeKleine, 1947), 472.
14. W. Stewart Woodfill re: G. Mennen Williams, August 10, 1975, Petersen Woodfill Correspondence, box 2, Mackinac Island State Park Commission (hereafter, MISPC) Archives, Keith R. Widder Library, MISPC Office, Mackinaw City, MI.
15. Rubin, *Bridging the Straits*, 9.
16. The bill passed the House 87 to 1 and was unanimously passed in the Senate. Rubin, *Bridging the Straits*, 11; *Journal of the Senate, 1950 Extra Session*, 319; *Journal of the House of Representatives of the State of Michigan, 1950 Extra Session* (Franklin DeKleine, 1950), 251.
17. *DFP*, July 21, 1958, 8.
18. *DFP*, December 29, 1959, 18. The cottage stood on state park land that had been leased by John Cudahy, who had built the building in 1888.
19. *Town Crier*, June 25, 1988, 1, 3. The coffin was brought to the island from St. Ignace,

where it was placed in a horse-drawn hearse and accompanied by an honor guard of uniformed military reenactors drawn from Fort Mackinac. At the interment ceremony, the reenactors fired a twenty-six-gun salute. The fort bugler blew taps to end the ceremony. Nearly three hundred individuals attended, including incumbent governor James Blanchard. All information taken from the *Town Crier* article of June 25, 1988.

20. Harvey Campbell Obituary, *DFP*, June 8, 1965, 8. Harvey Campbell to Lawrence Farrell, October 4, 1954, in folder, Mackinac Island Park Commission, 1954, box 136, G. Mennen Williams Papers, Bentley Historical Library, University of Michigan.
21. *DFP*, September 30, 1948, 3. In 1947, Sigler had not reappointed Doyle to the commission when Doyle's six-year term expired. In 1948, Sigler appointed Doyle to the commission for a term that would end in 1953. *DFP*, March 5, 1948, 11.
22. *DFP*, July 30, 1949, 2; August 23, 1949, 9.
23. MISPC minutes, August 29, 1949, 369, Keith R. Widder Library, MISPC Office, Mackinaw City, MI; David A. Armour, *100 Years at Mackinac: A Centennial History of the Mackinac Island State Park Commission, 1895–1995* (Mackinac Island State Historic Parks, 1995), 70.
24. Armour, *100 Years at Mackinac*, 70; *DFP*, July 30, 1949, 2; December 17, 1949, 19.
25. Armour, *100 Years at Mackinac*, 70; *DFP*, July 30, 1949, 1; August 23, 1949, 9.
26. MISPC minutes, December 16, 1949, 376, Keith R. Widder Library, MISPC Office, Mackinaw City, MI.
27. *DFP*, December 20, 1950, 15; MISPC minutes, December 27, 1950, 379, Keith R. Widder Library, MISPC Office, Mackinaw City, MI.
28. *DFP*, December 17, 1949, 19.
29. Lawrence A. Rubin, *Bridging the Straits*, 3.
30. In 1958, Williams and Doyle did, in fact, work together to pass a reorganization of state government that a commission chaired by Doyle had recommended five years earlier. Doyle was invited by Williams to the ceremonial signing of the legislation into law, where reporters noted that both he and Williams wore sheepish grins as Williams handed Doyle one of the pens used to sign the act into law. *DFP*, April 24, 1958, 38.
31. Doyle could work with, and for, Democratic governors. In 1940, for example, he supported Democrat Murray Van Wagoner for governor over his Republican opponent and seemed to have been quite friendly with him personally. Unpublished oral history of W. F. Doyle by Tom Downs, April 13, 1984, 21–24, Keith R. Widder Library, MISPC Office, Mackinaw City, MI.

32. *DFP*, February 22, 1951, 32.
33. *Oxford Leader*, October 5, 1951, 5.
34. *DFP*, April 3, 1953, 3; MISPC minutes, June 27, 1953, 403–404, Keith R. Widder Library, MISPC Office, Mackinaw City, MI.
35. *DFP*, April 1, 1953, 3.
36. Public Act 29, *Public and Local Acts of the Legislature of the State of Michigan Passed at the Regular Session of 1953* (Franklin De Kleine, 1953), 26.
37. *DFP*, August 24, 1955, 14.
38. *Town Crier*, August 12, 1955, 1.
39. G. Mennen Williams to James P. Dunnigan, September 6, 1955, MISPC, Boards & Commissions, 1955, box 165, William Papers, Bentley Library.
40. *DFP*, February 22, 1959, 25.
41. Memo to Lewis Beeson, September 25, 1956. Located in folder "Mackinac Island State Park Commission," file series "Boards & Commissions, 1956," box 197, William Papers, Bentley Library. The memo notes that the island government was "adamant" that the historic Clerk's Quarters of American Fur Company be torn down.
42. Charles Adrian to G. Mennen Williams, August 1, 1956, MISPC, Boards & Commission, 1956, box 197, Williams Papers, Bentley Library. The Beaumont Memorial is, more properly, the American Fur Company Retail Store. However the name of the site became permanently associated with Beaumont, because it was the place were he conducted groundbreaking experiments regarding human digestion.
43. W. Stewart Woodfill to Lawrence Farrell, July 30, 1957, box 227, folder 1957, Williams Papers, Bentley Library.
44. Lewis Beeson to W. F. Doyle, June 18, 1957. Mackinac Island State Park Commission, Boards & Commission, 1957, box 229, Williams Papers, Bentley Library. For Williams's position see Lawrence Farrell to G. Mennen Williams, March 26, 1957, also found in Mackinac Island State Park Commission, Boards & Commission, 1957, box 229, Williams Papers, Bentley Library, in which Farrell supports the historical commissions position and includes a handwritten note in Williams's trademark green ink and thus almost certainly written by him instructing Farrell to call Dunnigan and "handle" this. Apparently forgotten in this discussion was the fact that the Historical Commission and the Mackinac Island State Park Commission had cooperated in the early years of the twentieth century to erect historical markers on the island, including the Cass Memorial and

the marker at the Nicolet Watch Tower. However, even if the precedents from the 1910s had been remembered, it is not clear they would have been applied in the 1950s.

45. Armour, *100 Years at Mackinac*, 73; MISPC minutes, December 2, 1957, 465; January 7, 1958, 468; February 24, 1958, 482, Keith R. Widder Library, MISPC Office, Mackinaw City, MI.
46. Haskell, *Woodfill*, 141; *Town Crier*, June 2, 1984, 2.
47. *Town Crier*, July 8, 1955, 1.
48. Armour, *100 Years at Mackinac*, 74; Haskell, *Woodfill*, 140–142.
49. MISPC minutes, June 22, 1957, 426, Keith R. Widder Library, MISPC Office, Mackinaw City, MI.
50. Neil Staebler to G. Mennen Williams, May 17, 1957, General folder 2, Mackinac Island Park Commission, Boards & Commission, 1957, box 229, Williams Papers, Bentley Library.
51. W. Stewart Woodfill to James P. Dunnigan, June 18, 1957, General folder 2, Mackinac Island Park Commission, Boards & Commission, 1957, box 229, Williams Papers, Bentley Library.
52. Dunnigan to Woodfill June 2, 1957, General folder 2, Mackinac Island Park Commission, Boards & Commission, 1957, box 229, Williams Papers, Bentley Library.
53. Woodfill to Williams, June 24, 1957, General folder 2, Mackinac Island Park Commission, Boards & Commission, 1957, box 229, Williams Papers, Bentley Library.
54. MISPC minutes, August 12, 1957, 454–455, Keith R. Widder Library, MISPC Office, Mackinaw City, MI. Prior to the passage of the resolution, the commission had voted to abolish the understanding that the chair must be a member of the same political party as the sitting governor. Eugene Petersen inaccurately says "Dunnigan graciously gave up the chairmanship." Eugene T. Petersen, *Inside Mackinac* (Inside Mackinac, 1990), 26.
55. Regarding Dunnigan, Woodfill wrote to Eugene Petersen, "Mr. Dunnigan is a gentleman who wants everything he can possibly wrangle our [*sic*] of the Commission, but does not wish to give it, or do for it, anything!" W. Stewart Woodfill to Eugene Petersen, November 25, 1961, Petersen Woodfill Correspondence, box 1, MISPC Archives, Keith R. Widder Library, MISPC Office, Mackinaw City, MI. W. Stewart Woodfill to Eugene Petersen, November 8, 1960; Woodfill to "Confidential" November 2, 1960, Petersen Woodfill Correspondence,

box 1, folder Sept.–Dec. 1960, MISPC Archives, Keith R. Widder Library, MISPC Office, Mackinaw City, MI.

56. Petersen, *Inside Mackinac*, 27.
57. James P. Dunnigan, "Around My World in Eighty Years," unpublished memoir, December 1990, 31, Keith R. Widder Library, MISPC Office, Mackinaw City, MI.
58. MISPC minutes, August 12, 1957, 456.
59. *DFP*, December 16, 1958, 12. It is worth noting that at this meeting, the governor did not address the question of involving either Sault Ste. Marie or specifically talk about the city on Mackinac Island. He seems had learned, from his efforts in 1955, that he needed to scale back his ambitions and focus on what he could control—the park commission.
60. W. Stewart Woodfill, Statement re: G. Mennen Williams, August 10, 1975, notarized December 30, 1975, in Petersen Woodfill Correspondence, box 2, MISPC Archives, Keith R. Widder Library, MISPC Office, Mackinaw City, MI.
61. Petersen, *Inside Mackinac*, 5.
62. "1957 Annual Report of the Mackinac Island State Park Commission to the Governor," submitted December 2, 1957, folder 2, General, Boards and Commissions, 1957, box 29. Williams Papers, Bentley Library.
63. Armour, *100 Years at Mackinac*, 75.
64. Minutes of the Committee on Historic Restoration of Mackinac Island and the Straits Area, December 15, 1955, MISPC, Boards and Commissioners, 1957, box 165, Williams Papers, Bentley Library; Otto Lang to James Dunnigan, October 5, 1956, MISPC, Boards and Commissioners, 1957, box 165, Williams Papers, Bentley Library.
65. R. Daniel Musser Jr., "Summer 2012: An Oral History," conversation conducted by Phil Porter, director, Mackinac State Historic Parks (2021), 29, Keith R. Widder Library, MISPC Office, Mackinaw City, MI.
66. W. Stewart Woodfill to Eugene Petersen, April 15, 1961, Petersen Woodfill Correspondence, box 1, MISPC Archives, Woodfill, Keith R. Widder Library, MISPC Office, Mackinaw City, MI. Woodfill acknowledges his use of Doyle as the commission's "legislative agent" and admitted that for practical purposes, Doyle had come to serve as a commission member, with Woodfill's blessing. Woodfill to Petersen, April 15, 1961. On this subject, see also W. Stewart Woodfill to Commissioner Harry Teysen, September 3, 1966, Petersen Woodfill Correspondence, box 1, MISPC Archives, Keith R. Widder Library, MISPC Office, Mackinaw City, MI.

67. *DFP* February 2, 1958, 8; Armour, *100 Years at Mackinac*, 76.
68. The vote in the Senate was 22 to 6. *Journal of the Senate of the State of Michigan, 1958 Regular Session*, 2 vols. (Speaker-Hines and Thomas, 1958), 2:912. The bill passed the House 99 to 3. *Journal of the House of the State of Michigan, 1958 Regular Session*, 2 vols. (Speaker-Hines and Thomas, 1958), 2:1417. In separate legislation adopted in 1958, the legislature added a seventh commissioner to the Mackinac Island State Park Commission, representing Mackinaw City. For the text of the bill, see *Journal of the House, 1958*, 1:909. The House passed the bill unanimously. *Journal of the House, 1958*, 1:925. The Senate passed the bill 30 to 1. *Journal of the Senate, 1958*, 1:368–369.
69. Armour, *100 Years at Mackinac*, 77. Public Act 201, *Public and Local Acts of the Legislature of the State of Michigan Passed at the Regular Session of 1958* (Speaker-Hines and Thomas, 1958), 244–247.
70. MISPC minutes, April 28, 1958, 484–491, Keith R. Widder Library, MISPC Office, Mackinaw City, MI. Public Act 201, 244–247. Woodfill would much later note that despite the sum of $50,000 that is found in the resolution, he likely only actually advanced $6,200. W. Stewart Woodfill to Eugene Petersen, January 27, 1977, Petersen Woodfill Correspondence, box 2, MISPC Archives, Keith R. Widder Library, MISPC Office, Mackinaw City, MI.
71. Eugene T. Petersen, author's biographical statement in *Michilimackinac: History and Restoration*, rev. ed. (Mackinac Island State Park Commission, 1968).
72. Petersen, *Inside Mackinac*, 15–16.
73. *DFP*, June 14, 1959, 25.
74. *DFP*, June 19, 1958, 18; Armour, *100 Years at Mackinac*, 77, 113.
75. *DFP*, August 23, 1956, 48.
76. Petersen, *Inside Mackinac*, 73.
77. Woodfill, quoted in Petersen, *Inside Mackinac*, 73.
78. Petersen, *Inside Mackinac*, 35, 73.
79. MISPC minutes, April 24, 1954, 410, Keith R. Widder Library, MISPC Office, Mackinaw City, MI.
80. Armour, *100 Years at Mackinac*, 18; Petersen, *Inside Mackinac*, 38, 122.
81. Armour, *100 Years at Mackinac*, 77.
82. Eugene Petersen to W. Stewart Woodfill, November 17, 1965, Petersen to Woodfill Correspondence, box 1, MISPC Archives, Keith R. Widder Library, MISPC Office, Mackinaw City, MI.
83. Armour, *100 Years at Mackinac*, 77.

84. Attendance figures courtesy of Mackinac Island State Park Commission.
85. MISPC minutes, November 30, 1961, 725, Keith R. Widder Library, MISPC Office, Mackinaw City, MI.
86. MISPC minutes, November 30, 1961, 725.
87. Petersen, *Inside Mackinac*, 164, 112–113.

CHAPTER 7. THE DIRECTOR, THE MAYOR, AND HISTORY, 1961–1990

1. Eugene T. Petersen, *Inside Mackinac* (Inside Mackinac, 1990), 27.
2. W. Stewart Woodfill to Eugen Petersen, April 15, 1961, Petersen to Woodfill Correspondence, box 1, Mackinac Island State Park Commission (hereafter, MISPC) Archives, Keith R. Widder Library, MISPC, Mackinaw City, MI.
3. Petersen, *Inside Mackinac*, 130; *Michigan Manual, 1961–1962* (State of Michigan, 1961), 175, 184–186.
4. *Detroit Free Press* (hereafter, *DFP*), Mary 21, 1961, 17.
5. *Town Crier*, July 2, 1961, 8.
6. Petersen, *Inside Mackinac*, 128–130, 165.
7. Petersen, *Inside Mackinac*, 167.
8. Petersen, *Inside Mackinac*, 151–153.
9. Petersen, *Inside Mackinac*, 151–153.
10. Petersen, *Inside Mackinac*, 15–16.
11. *Town Crier*, July 30, 1967, 4.
12. *Town Crier*, June 15, 1969, 4. The relationship between Nordberg and Doyle during World War II is not documented, but given Doyle's one-man rule of the commission during those years, it seems it seems improbable that the person named superintendent of the park was not chosen personally by Doyle.
13. Petersen, *Inside Mackinac*, 42.
14. Petersen, *Inside Mackinac*, 153.
15. There is an extensive correspondence on this matter. See, for example, W. Stewart Woodfill to Eugene Petersen, May 28, 1966; May 30, 1966 (three letters); June 14, 1966 (2 letters); June 15, 1966; June 18, 1966; June 28, 1966; and July 1, 1966. Petersen writes to Woodfill May 21, 1966; May 30, 1966; June 4, 1966; and June 14, 196. All letters found in the Petersen/Woodfill Correspondence, box 2, folder May–August 1966, MSPC Archives, Keith R. Widder Library, MISPC Office, Mackinaw City, MI.

16. W. Stewart Woodfill to Eugene Petersen, September 7, 1967, Petersen Woodfill Correspondence, Box 2, MSPC Archives, Keith R. Widder Library, MISPC Office, Mackinaw City, MI.
17. *Town Crier*, July 10, 1966, 10.
18. MISPC minutes, October 18, 1967, 1072, Keith R. Widder Library, MISPC Office, Mackinaw City, MI.
19. *DFP*, October 22, 1967, 23.
20. *DFP*, October 22, 1967, 23.
21. MISPC minutes, June 7, 1968, 1109, Keith R. Widder Library, MISPC Office, Mackinaw City, MI.
22. MISPC minutes June 21, 1935, 247; February 17, 1936, 254; June 12, 1937, 265; June 7, 1968, 1110; Keith R. Widder Library, MISPC Office, Mackinaw City, MI.
23. W. Stewart Woodfill to William Doyle, May 6, 1968, copy found in Petersen Woodfill Correspondence, box 2, Keith R. Widder Library, MISPC Office, Mackinaw City, MI.
24. W. Stewart Woodfill to Eugene Petersen, May 4, 1968; W. Stewart Woodfill to George Romney, May 14, 1968; W. Stewart Woodfill to Eugene Petersen, May 19, 1968. All in Petersen Woodfill Correspondence, box 2, MISPC Archives, Keith R. Widder Library, MISPC Office, Mackinaw City, MI.
25. *Town Crier*, December 17, 1983, 25.
26. Woodfill to Petersen, May 4, 1968; Woodfill to Romney, May 14, 1968; Woodfill to Petersen May 19, 1968.
27. MISPC minutes, June 13, 1969, 1181–1182, Keith R. Widder Library, MISPC Office, Mackinaw City, MI.
28. *Town Crier*, August 11, 1973, 1, 7.
29. Eugene Petersen to W. Stewart Woodfill, February 2, 1970, Petersen and Woodfill Correspondence, box 2, MISPC Archives, MISPL, Keith R. Widder Library, MISPC Office, Mackinaw City, MI. Kelley seems to have eventually let the matter pass. As a member of the Mackinac Island State Park Commission from 1999 to 2015, he never raised the matter of the golf course lease, even though it can be argued that the lease remained below fair market value. The lease negotiated in 1925 set the rent at $1,500. This was the last lease negotiated prior to the Depression, when the commission's reduced the golf course rental fee to one dollar a year. Although the annual fee increased over time, it was not until 2024, when the annual golf course lease fee was set at $26,000, that the rent, adjusted for inflation, equaled that charged in 1925.
30. Attendance Statistics courtesy of Mackinac Island State Park Commission.

31. Petersen, *Inside Mackinac*, 112–113.
32. Department of the Army Corps of Engineers, *Waterborne Commerce of the United States: Part 3 Waterways and Harbors Great Lakes, Calendar Year 1961* (Bureau of the Census, US Department of Commerce, 1963), 157; *Waterborne Commerce of the United States: Part 3 Waterways and Harbors Great Lakes Calendar Year* (Water Resources Support Center, US Army Corps of Engineers, n.d.), 198, 9, 42.
33. Petersen, *Inside Mackinac*, 18.
34. *Town Crier*, June 18, 1967, 10; June 25, 1967, 2–3, Woodfill letter found on page 3.
35. *Town Crier*, June 25, 1967, 2.
36. *Town Crier*, August 25, 1968, 3.
37. *Town Crier*, July 28, 1962, 1.
38. *Town Crier*, June 29, 1974, 8. Regarding Coney Island, see Robin Jaffee Frank, *Coney Island: Visions of an American Dreamland, 1861–2008* (Yale University Press, 2015), or John F. Kasson, *Amusing the Million: Coney Island at the Turn of the Century* (Hill & Wang, 1978).
39. *Town Crier*, June 29, 1974, 8. Pfeiffelmann, an island entrepreneur, was capable of financing a Ferris wheel if he had wanted to build one, but there is no record that suggests he was serious.
40. *Town Crier*, June 29, 1974, 8.
41. *Town Crier*, June 7, 1975, 10.
42. Petersen, *Inside Mackinac*, 28.
43. *Town Crier*, June 7, 1975, 1, 10.
44. *Town Crier*, December 17, 1975, 17.
45. *Town Crier*, July 13, 1974, 3.
46. *Town Crier*, August 30, 1975, 3.
47. *Town Crier*, December 18, 1975, 17. Although small projects, defined as those costing less than $2,500, still needed a building permit, they were exempt from the requirement to submit plans that included the seal of an architect or engineer. *Town Crier*, July 20, 1974, 12.
48. *Town Crier*, June 21, 1975, 10; July 19, 1975, 12.
49. *Town Crier*, December 18, 1975, 17. The city's basic laws on this subject had last been comprehensively addressed in 1957. *Town Crier*, August 8, 1981, 2.
50. *Town Crier*, July 20, 1974, 4–5, 12.
51. *DFP*, February 5, 1979, 5.
52. McCabe, *Grand Hotel*, 222; *DFP*, June 3, 1979, 64.
53. *DFP*, June 11, 1979, 69.

54. Jack Matthews, review of *Somewhere in Time*, *DFP*, October 3, 1980, 17.
55. *Town Crier*, August 8, 1981, 2. *Town Crier*, June 13, 1981, 10.
56. *Town Crier*, August 6, 1988, 3.
57. *DFP*, January 6, 1980, 3, 5.
58. *DFP*, February 8, 1980, 3.
59. *DFP*, February 8, 1980, 15; February 26, 1980, 8.
60. *Town Crier*, August 13, 1977, 2.
61. *Town Crier*, July 2, 1977.
62. *Town Crier*, August 13, 1977, 23.
63. *Town Crier*, July 30, 1977, 4.
64. *Town Crier*, August 23, 1980, 7.
65. The history of the court battle over the tax, which at one point involved cases in four different courts, is told in *Town Crier*, June 10, 1978, 3; August 26, 1978, 3; December 16, 1978, 22; August 25, 1979, 4. The August 25, 1979, edition of the *Town Crier* reported that the newly formed Star Line, created by the purchase of Argosy Lines, was not a part of the lawsuit, which was being pursued by the Arnold and Shepler Ferry services. The *Town* Crier's coverage of this story continued on August 9, 1980, 2, 9; August 23, 1980, 7; September 6, 1980, 9; June 13, 1981, 8; September 5, 1981, 3; June 11, 1983, 8; July 9, 1983, 3; July 30, 1983, 3; September 10, 1983, 3; December 17, 1983, 11; June 23, 1984, 6; July 21, 1984, 3, 5; December 15, 1984, 7–8. The *Town Crier* issue of December 15, 1984, summarizes what the final agreement was between the ferry companies and the city government. The tax would be paid, but the major concessions were that in calculating the tax, revenues would not be included if the revenue was generated from freight traffic or derived from sources not directly related to the operation of the ferries, such as automobile parking fees.

CHAPTER 8. THE END OF AN ERA, 1961–1990

1. Marvin May, "Reminisces of Marvin May," 1999, 10, 21, Keith R. Widder Library, Mackinac Island State Park Commission (hereafter, MISPC) Office, Mackinaw City, MI. The sale agreement May signed had a ten-year noncompete clause, banning the Murdick's from making fudge. The Murdick's would return to the island fudge business in the 1950s, after the noncompete clause had expired. The new Murdick's fudge operation was sold in 1969 to the man who owned the business

next to their shop, Bob Benser Sr., "The Original Murdick's Fudge," https://www.originalmurdicksfudge.com/about-us/.

2. Phil Porter, *Mackinac: An Island Famous in These Regions* (Mackinac State Historic Parks, 1998), 68.
3. Joseph F. Davey, "Memories of Mackinac," unpublished manuscript, 1982, 5, Keith R. Widder Library, MISPC Office, Mackinaw City, MI.
4. For a brief history of the recurring conflicts between the carriage operators and the park commission, see David A. Armour, *100 Years Years at Mackinac: A Centennial History of the Mackinac Island State Park Commission, 1895–1995* (Mackinac Island State Historic Parks, 1995), 16–18. For Woodfill's conflicts during the 1950s with the group, see Eugene T. Petersen, *Inside Mackinac* (Inside Mackinac, 1990), 121–126.
5. "An Interview with R. D. Musser II," Grand Hotel, https://www.grandhotel.com/press-media/interview-r-d-musser-ii/.
6. *Town Crier*, August 18, 1963, 3. In making this statement Woodfill was splitting hairs. His evidence cited in the August 18, 1963, edition of the *Town Crier* is that the Grand Hotel had never paid a dividend to its owners. This is true, but as the hotel's sole owner, it simply meant Woodfill did not choose to pay himself a dividend.
7. John McCabe, *Grand Hotel Mackinac Island* (Lake Superior State College, 1987), 151.
8. *Town Crier*, August 18, 1963, 3. The minimum wage bill eventually became law in 1965. *Town Crier*, June 27, 1965, 8.
9. *Detroit Free Press* (hereafter, *DFP*), August 3, 1964, 4.
10. *DFP*, September 23, 1971, 1.
11. *DFP*, September 23, 1977, 1.
12. *DFP*, September 10, 1989, 1, 8.
13. *DFP*, September 14, 1989, 3; September 19, 1989, 4.
14. *DFP*, May 10, 1990, 66.
15. *DFP*, October 28, 1989, 30.
16. *DFP*, April 2, 1990, 17.
17. *DFP*, April 8, 1990, 29.
18. *DFP*, April 2, 1990, 17.
19. *DFP*, May 10, 1990, 66.
20. *DFP*, April 2, 1990, 17.
21. *DFP*, April 8, 1990, 29.
22. *DFP*, July 5, 1981, 33, 45.
23. *DFP*, November 27, 1976, 3.

24. *DFP*, June 19, 1977, 14; June 28, 1977, 3.
25. *DFP*, July 5, 1983, 33.
26. *DFP*, February 26, 1977, 6.
27. *DFP*, April 8, 1990, 29.
28. Phil Porter, *Fudge: Mackinac's Sweet Souvenir* (Mackinac State Historic Parks, 2001), 31.
29. *DFP*, November 24, 1969, 24.
30. *DFP*, September 19, 1965, 161.
31. *DFP*, September 19, 1965, 161; July 1, 1979, 92; November 24, 1969, 24. Harry Ryba's telling of the story about how he learned to make fudge from Murdick family members can be found in the *DFP* (July 1, 1979, 92) is different from the perspective of the Murdick family, explained by Phil Porter in *Fudge: Mackinac's Sweet Souvenir*, 31–32. Although the two tellings of the story differ in important details, what they agree on is that Ryba learned how to make fudge from descendants of Rome Murdick.
32. Porter, *Fudge*, 32.
33. *DFP*, September 19, 1965, 164.
34. *DFP*, September 19, 1965, 164.
35. *DFP*, September 19, 1965, 164; *Town Crier*, June 21, 1964, 5.
36. *DFP*, September 19, 1965, 166. Ryba was likely disappointed that, when President Gerald R. Ford visited the island in 1975, he stopped for his fudge at May's. *DFP*, June 14, 1975, 15.
37. *DFP*, September 19, 1965, 166.
38. *DFP*, September 19, 1965, 161; *Town Crier*, June 19, 1966, 11.
39. *DFP*, November 29, 1964, 24.
40. *DFP*, November 24, 1969, 24; July 1, 1979, 91; March 8, 1996, 18; *Town Crier*, July 28, 1968, 10. Technically Island House was leased by Ryba from the park commission, which owned it.
41. Gerald A. Ramin, "Tourism as a Major Resource of Mackinac Island" (Independent study, MS degree, Mankato State University, 1963), 49.
42. *Town Crier*, August 18, 1968, 3.
43. Cindy S. Aron, *Working at Play: A History of Vacations in the United States* (Oxford University Press, 1999), 238, 247. The advantages and disadvantages of paid vacations for industrial workers had been widely discussed during the 1920s, but in 1927, only about 5 percent of America's factory workers received a paid vacation. This changed radically in the late 1930s, with paid vacations often given

by manufacturers in the hopes of keeping workers from unionizing. See Aron, *Working at Play*, 248–249.

44. *DFP*, August 1, 1972, 22.
45. *DFP*, August 1, 1972, 22.
46. Eugene Petersen to W. Stewart Woodfill, July 21, 1974; Woodfill to Petersen, July 23, 1974. Both letters found in box 2, Petersen Woodfill Correspondence, MISPC Archives, Keith R. Widder Library, MISPC Office, Mackinaw City, MI. Woodfill was concerned about how those jealous of Ryba might cause problems, "it is likely some bastard will suggest to IRS that they look into Mr. Ryba's affairs," something Woodfill did not wish to have happen. Woodfill to Petersen, July 23, 1974.
47. *DFP*, September 19, 1965, 161, 163–164, 166. Who had the best fudge on the island was a subject of much controversy. In 1989, the *Detroit Free Press* created a panel of ten judges, who after tasting samples from the seven fudge makers then on the island ranked them this way: (1) Joann's, (2—a tie) May's and Murray Hotel, (4) Murdick's, (5) Ryba's, (6) Kilwins. *DFP*, August 27, 1989, 123.
48. *DFP*, July 1, 1979, 91, 92, 90.
49. *DFP*, July 1, 1979, 90.
50. *DFP*, November 24, 1969, 24.
51. Various dates appear in published sources for when Island House became state owned property. One newspaper gives the date as 1938. *Republican News and St. Ignace Enterprise*, January 9, 1947, 8. John Armour gives the date as 1942. Armour, *100 Years at Mackinac*, 60. As the date Armour gives is not footnoted, I have chosen to use the date printed in the *Republican News and St. Ignace Enterprise*.
52. Armour, *100 Years at Mackinac*, 60. The religious organization Moral Re-Armament (MRA) was a controversial organization that made friends and enemies around the globe and on Mackinac Island.
53. MISPC minutes, April 16, 1948, 344, 346–347, Keith R. Widder Library, MISPC Office, Mackinaw City, MI. *Republican News and St. Ignace Enterprise*, January 9, 1947, 8; January 30, 1947, 8; February 20, 1947, 8; April 3, 1947, 8; June 26, 1947, 5; May 20, 1948, 1; May 27, 1948, 8; August 5, 1948, 10; January 13, 1949, 8.
54. MISPC minutes, April 16, 1948, 347; December 29, 1948, 357; Keith R. Widder Library, MISPC Office, Mackinaw City, MI. *Republican New and St. Ignace Enterprise*, January 13, 1949, 8; June 30, 1949, 5.
55. *Town Crier*, July 8, 1972, 2. A circuit court order had closed the building in 1969 as a fire trap. *Town Crier*, July 8, 1972, 6.
56. *Town Crier*, August 27, 1971, 17; September 3, 1971, 13.

57. *Town Crier*, June 24, 1972, 5. *Town Crier*, June 18, 1983, 11.
58. *Town Crier*, June 23, 1973, 11.
59. *Town Crier*, August 27, 1971, 17; June 23, 1973, 10. The liquor license proved controversial. It required enabling legislation. This was the first state property to receive a liquor license. It also required action from the park commission, which under the law passed was technically the holder of the license. After the legislature passed the necessary law, an unexpected debate occurred. After "often heated discussion," the commission voted 4 to 3 to give Island House the license. *Town Crier*, July 28, 1972, 1. See also *Town Crier*, July 8, 1972, 2; July 14, 1972, 10; July 28, 1972, 1–3; June 23, 1973, 10.
60. Ryba also completely renovated the Lake View Hotel in 1983, another of his properties. Pamela A. Piljac and Thomas M. Piljac, *Mackinac Island: Historic Frontier, Vacation Resort, Timeless Wonderland* (Bryce-Waterton Publications, 1988), 159.
61. Armour, *100 Years at Mackinac*, 60.
62. MISPC minutes, December 29, 1948, 355–357, "Special Report on Island House," Keith R. Widder Library, MISPC Office, Mackinaw City, MI.
63. *DFP*, September 9, 1956, 28.
64. For a very sympathetic portrait regarding Buchman see H. W. "Bunny" Austin, *Frank Buchman as I Knew Him* (London: Grosvenor Books, 1975).
65. Otto Lang, *Town Crier*, September 3, 1977, 19.
66. *DFP*, April 1, 19, 1953, 3; September 19, 1955, 3; July 4, 1956, 3; September 9, 1956, 28; *Town Crier*, August 13, 1967, 1. See also W. F. Doyle , interview by Tom Downs, April 13, 1984, 3, Keith R. Widder Library, MISPC Office, Mackinaw City, MI.
67. Perhaps the most thoughtful review of the relationship between MRA and the people of Mackinac Island appeared in the *Town Crier*, fall edition, 1970, 2. Wrote the paper in an editorial, "They [MRA] seemed to want to transform the town of Mackinac Island to a model of their zeal for a puritan's haven."
68. *DFP*, March 30, 1966, 9; September 18, 1966, 60. It was reported in the *Town Crier*, July 20, 1969, 1, that the MRA sold the property to the college for $3.5 million.
69. *DFP*, September 30, 1969, 3; *Town Crier*, winter edition, 1969, 8.
70. *DFP*, April 6, 1971, 3.
71. *Town Crier*, June 25, 1971, 1; June 16, 1973, 4; *DFP*, April 10, 1971, 13.
72. *DFP*, February 14, 1973, 4; April 1, 1973, 12.
73. *Town Crier*, June 30, 1973, 4; *DFP*, July 2, 1973, 48. In its first incarnation as a commercial facility, it was known as the Mackinac College Inn, with 345 rooms.
74. The history of the property is documented in the *Town Crier* June 30, 1973, 4; July

14, 1973, 8; June 15, 1974, 11; June 26, 1976, 2; August 6, 1977, 2; September 11, 1982, 9.

75. George Cantor, *Explore Michigan: Mackinac* (University of Michigan, 2005), 29.
76. McCabe, *Grand Hotel*, 188, 194.
77. *Town Crier*, fall edition, 1975, 6. The city was also very much interested in working with the developers who had purchased the property as the City Planning Committee considered keeping the college campus together as a single parcel of land preferable to having it sold piecemeal to various buyers.
78. *DFP*, February 3, 1988, 6; June 15, 1988, 6; June 21, 1988, 48.
79. *DFP*, January 17, 1988, 6.
80. James P. Dunnigan, "Around My World in Eighty Years," unpublished memoir, December 1990. Keith R. Widder Library, MISPC Office, Mackinaw City, MI.
81. *DFP*, July 24, 1996, 16.
82. Petersen, *Inside Mackinac*, 204.
83. *DFP*, August 27, 1989, 118.
84. *DFP*, Mach 8, 1996, 18.
85. *DFP*, April 21, 1987, 3, 10.
86. *DFP*, July 14, 1975, 5.
87. Petersen, *Inside Mackinac*, 202.
88. Petersen, *Inside Mackinac*, 202.
89. R. Daniel Musser Jr., "Summer 2012: An Oral History," conversation conducted by Phil Porter, director, Mackinac State Historic Parks (2021), 19, Keith R. Widder Library, MISPC Office, Mackinaw City, MI.
90. Musser, "Oral History," 4.
91. McCabe, *Grand Hotel*, 169; Musser, "Oral History," 21.
92. McCabe, *Grand Hotel*, 169, 185.
93. *DFP*, August 1, 1972, 22.
94. Musser, "Oral History," 28.
95. For a typical story, this one on renovations to the hotel's main lobby and the relocation of the registration desk to a new level within the hotel, see *DFP*, July 16, 1978, 21. Some, such as Petersen, did question the wisdom of allowing an Eastern designer a free hand in renovating a classic Midwestern summer resort. Petersen, *Inside Mackinac*, 211.
96. McCabe, *Grand Hotel*, 195, 198, 222.
97. *Town Crier*, June 10, 1978, 2, 7.
98. H. Morgan Haskell, *W. Stewart Woodfill: Master of Mackinac's Grand Hotel* (H. Morgan Haskell, 2001), 179.

99. McCabe, *Grand Hotel*, 202; Haskell, *Woodfill*, 181–182. Woodfill owned the hotel through a trust. He thus called the trustee into his office and dictated a sales contract. The trustee in turn carried the document to Musser, who, in the presence of a witness, accepted it as delivered. Woodfill was not wrong in thinking that Musser's plans would be very expensive. It was reported that the structural and redecorating projects cost $12 million to complete. *DFP*, July 5, 1987, 10.
100. McCabe, *Grand Hotel*, 203. See, for example, *DFP*, August 9, 1979, which was reporting that Woodfill had given the hotel to Musser.
101. McCabe, *Grand Hotel*, 203; W. S. Woodfill Obituary, *New York Times*, March 7, 1984, B4.
102. Haskell, *Woodfill*, 182.
103. *DFP*, August 9, 1979, 10.
104. Musser, "Oral History," 29.
105. Musser, "Oral History." McCabe, *Grand Hotel*, 204–205; Haskell, *Woodfill*, 162; Musser, "Oral History," 29. Pellston was the nearest airport with commercial airline service.
106. McCabe, *Grand Hotel*, 228; W. S. Woodfill Obituary, *New York Times*, March 7, 1984, B4.

CHAPTER 9. THE BEGINNING, 1895–1932

1. Brian Leigh Dunnigan, *A Picturesque Situation: Mackinac before Photography, 1616–1860* (Wayne State University, 2008), 191–193; Russell McKee, ed., *Mackinac: The Gathering Place* (Michigan Natural Resources Magazine, 1981), 82.
2. *Census of the State of Michigan, 1874* (Lansing, 1875), tables, 28.
3. Emerson R. Smith, comp. *Before the Bridge: A History and a Directory of St. Ignace and Nearby Localities* (Kiwanis Club of St. Ignace, 1957), 7, 6, 79; St. Ignace Public Library, *St. Ignace* (Arcadia, 2008), 7, 21; "Fort Michilimackinac," miplace, Michigan Economic Development Corporation, https://www.miplace.org/historic-preservation/programs-and-services/historic-landmarks/fort-michilimackinac/. The exact date the French established Fort de Buade is subject to discussion. For example, the St. Ignace Public Library staff dated the fort's founding as 1688. St. Ignace Public Library, *St. Ignace*, 7.
4. Graydon M. Meints, *The Fishing Line: A History of the Grand Rapids & Indiana Railroad* (Michigan State University Press, 2018), 79.

5. St. Ignace Public Library, *St. Ignace*, 7, 30, 52–54. The Martel Furnace, which opened in 1881, would operate until 1894. It would struggle on for another decade, operating intermittently, before being destroyed by fire in 1903. Information about the Martel Furnace from St. Ignace Public Library, *St. Ignace* (Arcadia, 2008), 30.
6. *St. Ignace Enterprise*, November 27, 1913, 1, 4.
7. *Michigan Farming Lands!*, broadside issued by the Detroit, Mackinac & Marquette Railroad, Co., W.O. Strong, land commissioner, January 30, 1882, Clarke Historical Library, Central Michigan University.
8. Aaron Shapiro, *The Lure of the North Woods: Cultivating Tourism in the Upper Midwest* (University of Minnesota Press, 2013), 1. Canadians were of the same opinion. See Andrew Watson, *Making Muskoka Tourism, Rural Identity, and Sustainability* (University of British Columbia Press, 2022), 20.
9. *St. Ignace Enterprise*, May 28, 1914, 1.
10. *St. Ignace Enterprise*, November 27, 1913, 1.
11. Watson, *Making Muskoka*, 41, 94. By 1895, Muskoka had about four hundred summer cottages where the occupants resided for the season, and the summer population was about the same size as the number of people who lived year-round in the six townships that made up the Muskoka region. Watson, *Making Muskoka*, 82.
12. *St. Ignace Enterprise*, December 1, 1913, 1.
13. *St. Ignace Enterprise*, March 16, 1916, 1.
14. *St. Ignace: The Gate-Way City to the Upper Peninsula* (Business Men's Association, 1901).
15. St. Ignace Public Library, *St. Ignace*, 50–51. *St. Ignace Enterprise*, May 14, 1914, 1.
16. Cindy S. Aron, *Working at Play: A History of Vacations in the United States* (Oxford University Press, 1999), 111–126.
17. *St. Ignace Enterprise*, January 21, 1915, 1.
18. *St. Ignace Enterprise*, March 16, 1916, 1.
19. *St. Ignace Enterprise*, August 10, 1916, 1.
20. As noted in St. Ignace Public Library, *St. Ignace*, 24, lumbering did not end in the 1920s. It continued in the city until the 1950s, although on a drastically reduced scale.
21. *Official Directory and Legislative Manual of the State of Michigan, 1891–1892* (Lansing, 1891), 649; *Michigan Official Directory and Legislative Manual, 1923–1924* (n.p., n.d.), 228.
22. Shapiro, *Lure of the North Wood*, 11.

23. *St. Ignace Enterprise*, May 11, 1922, 1.
24. *St. Ignace Enterprise*, May 11, 1922, 1.
25. William Henry Barett, "The History of Transportation across the Straits of Mackinac" (MA Thesis, University of Hawaii, 1963), in *Memories of Mackinaw*, comp. Judy Ranville and Nancy Campbell (Little Traverse Printing, 1976), 130.
26. Frank F. Rogers, *History of the Michigan State Highway Department, 1905–1933* (Franklin DeKleine, 1933), 144.
27. US Census, 1920 Decennial Census, MI. Census data by state and county for 1920 can be found online at https://www2.census.gov/library/publications/decennial/1920/volume-1/41084484v1ch2.pdf.
28. Barrett, "The History of Transportation across the Straits of Mackinac," 129.
29. *Twenty-First Biennial Report of the State Highway Commissioner for the Fiscal Years Ending June 30, 1945 and June 30, 1946* (Franklin DeKleine, 1946), 161.
30. *Twenty-First Biennial Report of the State Highway Commissioner*, 161. *Report of the Chief of Engineers, U.S. Army*, part 2, *1930* (Government Printing Office 1930), 843.
31. *St. Ignace Enterprise*, July 6, 1922; April 23, 1925; April 30, 1925.
32. *St. Ignace Enterprise*, August 13, 1925, 1.
33. *St. Ignace Enterprise*, July 5, 1923, 1; April 22, 1926, 1.
34. *St. Ignace Enterprise*, August 30, 1923, 1; October 4, 1923, 1.
35. *St. Ignace Enterprise*, July 5, 1923, 1. The Martel Furnace had been constructed in St. Ignace to make iron from Upper Peninsula iron ore but was relatively short lived. *St. Ignace Enterprise*, January 30, 1930, 1.
36. *St. Ignace Enterprise*, June 2, 1927; 1; April 18, 1928, 1. Evergreen Shores also featured a restaurant.
37. *St. Ignace Enterprise*, June 4, 1925, 5.
38. St. Ignace Public Library, *St. Ignace*, 75. Madeline Okerman Adie, *The Straits of Mackinac* (Arcadia, 2012), 45. Fort Algonquin housed a store selling tourist-related items.
39. *St. Ignace Enterprise*, July 13, 1922, 1.
40. *St. Ignace Enterprise*, April 21, 1927, 1.
41. *St. Ignace Enterprise*, July 14, 1927, 1; July 21, 1927, 1.
42. *St. Ignace Enterprise*, August 4, 1927, 1.
43. *St. Ignace Enterprise*, July 14, 1927, 1; July 21, 1927, 1.
44. *Bay City Tribune* as reprinted in the *St. Ignace Enterprise*, August 27, 1925, 1.
45. *Cheboygan Democrat* as reprinted in the *St. Ignace Enterprise*, August 23, 1923, 1.
46. *St. Ignace Enterprise*, August 23, 1923, 1.

47. *St. Ignace Enterprise*, September 13, 1923, 1.
48. *St. Ignace Enterprise*, October 4, 1923, 1.
49. *St. Ignace Enterprise*, December 20, 1923, 1.
50. *Republican News and St. Ignace Enterprise*, February 6, 1936, 1.
51. *St. Ignace Enterprise*, August 29, 1929, 1.
52. *St. Ignace Enterprise*, July 11, 1929, 1.
53. *St. Ignace Enterprise*, April 23, 1925, 1; July 22, 1926, 1; November 1, 1928, 1.
54. *St. Ignace Enterprise*, September 8, 1932, 1.
55. "The Traverse Region," in Ranville and Campbell, *Memories of Mackinaw*, 9–10.
56. Barrett, "History of Transportation across the Straits," 121.
57. Matthew J. Friday, *Among the Sturdy Pioneers: The Birth of the Cheboygan Area as a Lumbering Community, 1778–1935* (Trafford, 2006), 40.
58. Barrett, "History of Transportation across the Straits of Mackinac," 124.
59. Meints, *Fishing Line*, 78.
60. Friday, *Among the Sturdy Pioneers*, 41.
61. Meints, *Fishing Line*, 78.
62. Barrett, "History of Transportation across the Straits," 128.
63. Adie, *Mackinaw City*, 8.
64. Meints, *Fishing Line*, 80, 108.
65. Mary Desy, "The Desy Family," in Ranville and Campbell, *Memories of Mackinaw*, 237.
66. Leona Brown, "Mackinaw City in the Old Days about 1908," in Ranville and Campbell, *Memories of Mackinaw*, 227.
67. M. Christine Byron and Thomas R. Wilson, *Vintage Views along the West Michigan Pike: From Sand Trails to US-31* (Arbutus Press, 2011), 14; Lena E Wallin, interview by Judy La Way Ranville, in Ranville and Campbell, *Memories of Mackinaw*, 266.
68. *Town Crier*, December 1973, 33.
69. Barrett, "The History of Transportation across the Straits of Mackinac," 123.
70. P. Donan, "Michigan Central Tourist Guide, Mackinac Island, 1883," in Ranville and Campbell, *Memories of Mackinaw*, 112.
71. *Cheboygan Democrat*, July 1, 1910, 4.
72. *Detroit Free Press* (hereafter, *DFP*), June 28, 1914, C6.
73. Wa-Wa-Tam Association, "Wawatam Memories" (1958), in Ranville and Campbell, *Memories of Mackinaw*, 69, 74; R. L. Polk and Co., *Michigan State Gazateer and Business Directory, 1907–1908* (R. L. Polk, 1907), 1387.
74. M. Christine Byron and Thomas R. Wilson, *Vintage Views of the Mackinac Straits*

Region (Arbutus Press, 2007), 13.

75. John Moore, "A Reporter at Large: A Little Journey to Mackinaw," in Ranville and Campbell, *Memories of Mackinaw*, 79.
76. Maire Patrick (Riggs), "Memories of Yesteryear," in Ranville and Campbell, *Memories of Mackinaw*, 231.
77. *DFP*, May 31, 1913, 3.
78. *DFP*, November 14, 1915, 10.
79. Moore, "A Reporter at Large," 80.
80. Brown, "Mackinaw City in the Old Days," 228.
81. Marne Smith, "Reminisce," 4, 8, Michigan History Room, Mackinaw Area Public Library, Mackinaw City.
82. Adie, *Mackinaw City*, 8.
83. Steven R. Campbell, "The Development of Mackinaw City," in Ranville and Campbell, *Memories of Mackinaw*, 52; Edna Coffman, *Mackinaw City Settlers and the Savage Straits* (Voyager Press, 1976), 19, 26. How much harvestable timber was near Mackinaw City is not certain. Although most sources, cited earlier, seem to think it was relatively limited, Kathryn Washburn Breighner, "A Tale of Two Cities," in *Mackinac: The Gathering Place*, ed. Russell McKee ([Michigan Natural Resources Magazine, 1981], 86) writes that there was 150 million board feet of lumber growing within "working distance" of Mackinaw City and "around the turn of the century" the community was home to several mills and a clothespin factory. This may be true, but in the end, Cheboygan, not Mackinaw City, was the principal lumbering location in the Straits region of Michigan's Lower Peninsula.
84. Judy Ranville, "Ralph Preston's Railroad Memories of 1924," in Ranville and Campbell, *Memories of Mackinaw*, 110.
85. Ken Teysen, *Cheboygan Daily Tribune*, June 21–23, 1996, in "Mackinaw City MI & Vicinity Residents" notebook, in Michigan History Room, Mackinaw Area Public Library.
86. Smith, "Reminisce," 8.
87. Grand Rapids and Indiana Railway, *Annual Report* (1915), 16.
88. Barrett, "History of Transportation across the Straits," 135, 133.
89. *State of Michigan Twelfth Biennial Report of the State Highway Commissioner for the Fiscal Years June 30, 1927 and June 30, 1928* (Franklin DeKleine, 1929), 53.
90. *State of Michigan Twelfth Biennial Report of the State Highway Commissioner for the Fiscal years June 30, 1927 and June 30, 1928* (Franklin DeKleine, 1929), 53.

CHAPTER 10. AUTOMOBILE TOURISM, 1912–1959

1. *Emmet County Graphic*, September 10, 1925, as quoted in M. Christine Byron and Thomas R. Wilson, *Vintage Views of the Mackinac Straits Region* (Arbutus Press, 2007), 19.
2. *Petoskey Evening News*, July 18, 1929, as quoted in in Byron and Wilson, *Vintage Views of the Mackinac Straits*, 20.
3. Theodore Dreiser as quoted in Warren James Belasco, *Americans on the Road: From Autocamp to Motel, 1910–1945* (MIT Press, 1979), 19.
4. John A. Jakle, *The Tourist: Travel in Twentieth-Century North America* (University of Nebraska Press, 1985), 121, 123.
5. Bill Kulsea and Tom Shawver, *Making Michigan Move: A History of Michigan Highways and the Michigan Department of Transportation* (Michigan Department of Transportation, 1992), 5, 8.
6. Jakle, *Tourist*, 125–126.
7. Frank F. Rogers, *History of the Michigan State Highway Department, 1905–1933* (Franklin DeKleine, 1933), 166.
8. M. Christine Byron and Thomas R. Wilson, *Vintage Views along the West Michigan Pike: From Sand Trails to US-31* (Arbutus Press, 2011), 23.
9. "M-180," Michigan Highways, http://www.michiganhighways.org/listings/M-108.html. In Michigan, the Dixie Highway had two branches, one of which came from Indianapolis to connect to the West Michigan Pike and the other of which came north from Toledo to Detroit, and then northward along the East Michigan Pike. See the highway route map published in *The Dixie Highway* 10, no. 6 (July 1924): 12.
10. Jakle, *Tourist*, 152–153.
11. *Journal of the Senate of the State of Michigan, 1903*, 2 vols. (Robert Smith, 1903), 2:1462–1463; David A. Armour, *100 Years at Mackinac: A Centennial History of the Mackinac Island State Park Commission, 1895–1995* (Mackinac Island State Historic Parks, 1995), 34, 88. The *Journal of the Senate* dates the state's acceptance of the property to 1903. Armour (*100 Years at Mackinac*, 88) says the state accepted the park from the city in 1904. The difference may reflect when the legislature agreed to accept the land and when the actual paperwork was completed. Since Armour did not include a footnote to document the 1904 date, I have chosen to use the year when the legislature passed the necessary act.
12. Armour, *100 Years at Mackinac*, 48, 88. *Grand Rapids Herald*, August 15, 1920, as quoted Byron and Wilson, *Vintage Views of the Mackinac Straits Region*, 36.

13. Armour, *100 Years at Mackinac*, 48, 8; Judy Ranville, "Chris Schneider and His Park," in *Memoires of Mackinaw*, comp. Judy Ranville and Nancy Campbell (Little Traverse Printing, 1976), 84; *Detroit Free Press* (hereafter, *DFP*), July 13, 1919, B8.
14. Aaron Shapiro, *The Lure of the North Woods: Cultivating Tourism in the Upper Midwest* (University of Minnesota Press, 2013), 22–24; Jakle, *Tourist*, 159; Belasco, *Americans on the Road*, 118–119.
15. Ranville, "Chris Schneider and His Park," 84; Armour, *100 Years at Mackinac*, 68.
16. Belasco, *Americans on the Road*, 137.
17. Armour, *100 Years at Mackinac*, 48, 42.
18. Jakle, *Tourist*, 164, 163.
19. *Petoskey Evening News*, June 12, 1930, as quoted in Byron and Wilson, *Vintage Views of the Mackinac Straits Region*, 25.
20. Byron and Wilson, *Vintage Views of the Mackinac Straits Region*, 26.
21. Peter Varney, *Motor Camping* (Leisure League of America, 1935), 38.
22. *DFP*, October 10, 1937, 2; *Republican News and St. Ignace Enterprise*, June 13, 1935, 1; August 20, 1936, 1; January 5, 1939, 6.
23. Teysen's 50th anniversary brochure, located in unlabeled binder of newspaper clippings and other material, Michigan History Room, Mackinaw Area Public Library.
24. H. J. Teysen obituary, *Petoskey News Review*, February 2, 1960, 1; Roger Andrews, "Frontier Restaurant: A Teysen Tradition," originally published in the *Mackinac Island News*, July 26, 1941, in Ranville and Campbell, *Memories of Mackinaw*, 301; Madeline Okerman Adie, *Mackinaw City* (Arcadia, 2007), 58.
25. Andrews, "Frontier Restaurant," 301–302.
26. Jakle, *The Tourist*, 167, 168.
27. Teysen's slogan is found in an anonymous clipping reprinting an article first published in the *Menominee Herald Leader*, September 11, 1935. The anonymous clipping is found in an unlabeled binder of newspaper clippings housed in the Michigan History Room, Mackinaw Area Public Library.
28. Andrews, "Frontier Restaurant," 301.
29. Marne Smith, "My Life Growing Up in a Small Town [Mackinaw City]," unpublished memoir, 1980, Michigan History Room, Mackinaw Area Public Library, 4–5.
30. Byron and Wilson, *Vintage Views along the West Michigan Pike*, 32.
31. Belasco, *Americans on the Road*, 142–143, 150, quotation at 143.
32. Teysen's 50th anniversary brochure, 1976.
33. Mackinac Island State Park Commission, *1937–1938 Report of the Mackinac Island*

State Park Commission of Michigan (1938), 23.

34. Armour, *100 Years at Mackinac*, 89; *DFP*, June 25, 1933, 10.
35. Armour, *100 Years at Mackinac*, 110. In 1913, the commission had authorized a careful survey of the site (68).
36. *DFP*, March 4, 1934, 8; April 15, 1935, 8; September 26, 1936; *Republican News and St. Ignace Enterprise*, May 27, 1933, 1.
37. *DFP*, June 18, 1933, 13.
38. "Annual Re-enactment Pontiac Massacre 1763," flyer in Ranville and Campbell, *Memories of Mackinaw*, 91. "Official Program, Opening Old Fort Michilimackinac Stockade, Mackinaw City, July 1st, 1933," in Ranville and Campbell, *Memories of Mackinaw*, 93.
39. *DFP*, December 7, 1933, 8.
40. *DFP*, March 4, 1934, 8.
41. Mackinac Island State Park minutes, September 27, 1935, 250, Keith R. Widder Library, Mackinac Island State Park Commission Office, Mackinaw City, MI. The commission appears to have had little interest in working with St. Ignace. In 1938, St. Ignace's city government tried to give the community's troubled golf course to the Mackinac Island Park Commission. Although the commission had some initial interest, it eventually declined involvement in the project, again pleading "insufficient funds." *Republican News and St. Ignace Enterprise*, June 2, 1938, 1; June 30, 1938, 1; August 10, 1939, 1; August 1, 1940, 1. The golf course would eventually become owned by the county government. *Republican News and St. Ignace Enterprise*, October 24, 1940, 1. It should be added that the commission had received some land in St. Ignace in 1923; however the land was later given by the commission to the state's Department of Conservation. Armour, *100 Years*, 100.
42. St. Ignace may have made an effort to have the state legislature require the Mackinac Island Park Commission to become involved in the fort project. The park commission was not necessarily opposed to the idea, although likely with its usual demand that it receive additional state funds for the new project. The idea never seems to have gained serious support within the legislature. See Mackinac Island State Park Commission, *1933 and 1934 Report of the Mackinac Island State Park Commission*, report submitted December 1, 1934, 44.
43. *St. Ignace Enterprise*, September 18, 1930, 1.
44. *St. Ignace Enterprise*, March 5, 1931, 1.
45. On the location of Fort De Baude, see Eugene Petersen's comments made at a public meeting in St. Ignace. Eugene Petersen, *Inside Mackinac* (Inside Mackinac,

1990), 53.

46. *Republican News and St. Ignace Enterprise*, August 24, 1933, 1. *Republican News and St. Ignace Enterprise*, July 19, 1934, 1; July 26, 1934, 1; August 9, 1934, 1.
47. *Republican News and St. Ignace Enterprise*, August 16, 1934, 1.
48. *Republican News and St. Ignace Enterprise*, August 13, 1936, 1. Father Howard said that he had not found an illustration of Fort de Baude, but he had discovered enough documents regarding French frontier forts of the period to know what the fort would have looked like. Father Howard's work raises the question of what model for fort reconstruction had been proposed in 1934. Another call to rebuild the fort appeared in the paper December 31, 1936, 1.
49. *Republican News and St. Ignace Enterprise*, January 21, 1937, 1.
50. *Republican News and St. Ignace Enterprise*, August 5, 1937, 1; August 19, 1937, 1.
51. *Republican News and St. Ignace Enterprise*, September 2, 1937, 1.
52. *St. Ignace Enterprise*, January 16, 1930, 1.
53. *St. Ignace Enterprise*, January 30, 1930, 1.
54. *St. Ignace Enterprise*, January 16, 1930, 1; January 30, 1930, 1.
55. As quoted in the *St. Ignace Enterprise*, February 13, 1930, 1.
56. *St. Ignace Enterprise*, January 30, 1930, 1; February 13, 1930, 1.
57. *Republican News and St. Ignace Enterprise*, March 7, 1940, 1; March 21, 1940, 1.
58. *Republican News and St. Ignace Enterprise*, March 7, 1940, 1.
59. *Republican News and St. Ignace Enterprise*, January 9, 1941, 1.
60. *Republican News and St. Ignace Enterprise*, January 9, 1941, 1.
61. *Republican News and St. Ignace Enterprise*, August 27, 1942, 1.
62. *DFP*, January 15, 1938, 5.
63. *DFP*, August 1, 1943, 45.
64. Cindy S. Aron, *Working at Play: A History of Vacations in the United States* (Oxford University Press, 1999), 238, 247. The advantages and disadvantages of paid vacations for industrial workers had been widely discussed during the 1920s, but in 1927 only about 5 percent of America's factory workers received a paid vacation. This changed radically in the late 1930s, with paid vacations often given by manufacturers in the hopes of keeping workers from unionizing. See Aron, *Working at Play*, 248–249.
65. Belasco, *Americans on the Road*, 170.
66. Harvey H. Jackson III, *Redneck Riveria: An Insider's History of the Florida-Alabama Coast* (University of Georgia Press, 2012), 40.
67. *DFP*, May 26, 1947, 25.

68. *DFP*, May 26, 1947, 25; Armour, *100 Years at Mackinac*, 68.
69. *DFP*, June 6, 1948, 6.
70. *Republican News and St. Ignace Enterprise*, April 10, 1947, 1.
71. *Republican News and St. Ignace Enterprise*, April 17, 1947, 1.
72. *Oxford Leader*, August 6, 1948, 8. Clarke Historical Library, Central Michigan University, Digital Michigan Newspaper Database, https://digmichnews.cmich.edu/.
73. *Republican News and St. Ignace Enterprise*, March 13, 1947, 1.
74. *DFP*, February 20, 1954, 8.
75. *DFP*, November 1, 1957, 3.
76. *DFP*, June 28, 1953, 3.
77. *DFP*, November 2, 1957, 3.
78. *DFP*, November 2, 1957. 3.
79. *Republican News and St. Ignace Enterprise*, July 19, 1956, 1.
80. *Republican News and St. Ignace Enterprise*, February 6, 1958, 1.
81. *Republican News and St. Ignace Enterprise*, February 20, 1958, 1.
82. Press Release, Mackinaw City Chamber of Commerce, reprinted in the *Mackinaw News*, February 13, 2022, 9.
83. *DFP*, July 3, 1955, 19.
84. Presentation of Greg Teysen to the Mackinaw Area Historical Society, January 8, 2024, Mackinaw Area Public Library, Mackinaw City, MI.
85. *DFP*, November 21, 1953, 4.
86. *DFP*, May 8, 1954, 3; May 9, 1954, 3, 19; *Saline Observer* April 22, 1954, 5; Armour, *100 Years at Mackinac*, 80.
87. *DFP*, November 2, 1957, 1.
88. Mike Fornes, *Mackinac Bridge: A 50 Year Chronicle* (Cheboygan Tribune, 2007), 23.
89. *DFP*, November 2, 1957, 1.
90. *DFP*, October 20, 1957, 60.
91. *DFP*, October 27, 1957, 3.
92. *DFP*, November 2, 1957, 3; Fornes, *Mackinac Bridge*, 28.
93. Fornes, *Mackinac Bridge*, 29.
94. *DFP*, June 26, 1958, 3; June 27, 1958, 1; June 28, 1958, 2; June 29, 1958, 1.
95. *DFP*, May 5, 1958, 40; Fornes, *Mackinac Bridge*, 29–32.
96. *DFP*, June 26, 1958, 3; June 27, 1958, 1; June 28, 1958, 2; June 29, 1958, 1.
97. *DFP*, June 27, 1958, 1.
98. *DFP*, June 29, 1958, 1.

99. *DFP*, June 29, 1958, 11.
100. *DFP*, June 26, 1958, 3.
101. *DFP*, June 28, 1958, 4.

CHAPTER 11. THE MORE THINGS CHANGE, 1960–1990

1. *Detroit Free Press* (hereafter, *DFP*), April 24, 1960, 21.
2. *DFP*, June 30, 1963, 34.
3. *DFP*, June 30, 1963, 34.
4. US Department of Commerce, Bureau of the Census, *1960 Census of the Population: Advance Reports, Final Population Count, Michigan, November 30, 1960* (PC[A1]-24), table 2, pages 6–7, 10.
5. Paul Stemkowski, "The Fort Gift Shop & Candy Shop," presentation made on September 9, 2024, at the Mackinaw Area Historical Society, Mackinaw City, MI.
6. *DFP*, June 30, 1963, 34. I-75 was part of the new interstate highway system built in the 1960s and 1970s. In Michigan, it ran from the Ohio border, through Detroit, and then northward to the Mackinac Bridge, from where it continued to Sault St. Marie. I-75 was completed in Michigan in 1962. *DFP*, June 10, 1962. The new interstate highway did, however, cause controversy among some Mackinac Island business owners, who complained that while there were exit signs clearly noting both Mackinaw City and St. Ignace, there was no exit sign mentioning Mackinac Island. Responding state officials chose not to comment on the irony of an island community that did not permit automobiles demanding a freeway exit sign. Instead, they blamed the lack of signage on federal regulations. *Town Crier*, August 12, 1962, 8. Island merchants, however, were not to be put off and continued the fight. Eventually, an exasperated state highway commissioner pointed out that every intermediate town that did not appear on an exit sign would likely want its name on one. He added that the island's tourist trade was doing quite well without a freeway exit to call its own. *Town Crier*, August 11, 1963, 1.
7. *DFP*, June 30, 1963, 34.
8. *DFP*, March 3, 1964, 15.
9. Eugene T. Petersen, *Inside Mackinac* (Inside Mackinac, 1990), 51–52; Mackinac Island State Park Commission (hereafter, MISPC) minutes, December 2, 1957, 465, Keith R. Widder Library, MISPC Office, Mackinaw City, MI.
10. Petersen, *Inside Mackinac*, 52.

11. Petersen, *Inside Mackinac*, 53. Underscoring the problem of determining where the fort had stood was an offer by the state, made in 1956, to place a historical marker on the site of Fort de Baude. The project ran into the old problem of where the fort had been located. See *Republican News and St. Ignace Enterprise*, October 23, 1956, 1.
12. Petersen, *Inside Mackinac*, 53.
13. Petersen, *Inside Mackinac*, 54.
14. Petersen, *Inside Mackinac*, 54.
15. Petersen, *Inside Mackinac*, 27, 54–55.
16. MISPC minutes, December 1, 1962, 801, Keith R. Widder Library, MISPC Office, Mackinaw City, MI.
17. *Town Crier*, August 27, 1983, 2; December 17, 1983, 25.
18. *Town Crier*, December 17, 1983, 25.
19. *Town Crier*, August 27, 1983, 14.
20. December 17, 1983, 25. The *Town Crier* covered the hearings when they occurred in their August 27 edition and also offered a long summary of the hearings, including many quotes from speakers at the hearings that had not been previously published in the paper's end of year edition, published December 17.
21. *Town Crier*, August 27, 1983, 14; December 17, 1983, 25.
22. *Town Crier*, August 27, 1983, 15.
23. *Town Crier*, August 16, 1964, 11; June 20, 1965, 3.
24. *Town Crier*, August 4, 1984, 3.
25. MISPC minutes, July 26, 1986, 2185; August 16, 1986, 2189, Keith R. Widder Library, MISPC Office, Mackinaw City, MI; "Bridging the Straits: A Summary Report Concerning the Feasibility of a Proposed Upper Peninsula Welcome Center/ Straits of Mackinac Transportation Museum," presented by the Mackinac Island State Park Commission, August 15, 1988, RG-98, Michigan Department of Transportation, Travel Information and Welcome Centers, box 6, folder "Upper Peninsula," folder 2 of 2, archives of Michigan, Lansing, MI.
26. *Cheboygan Daily Tribune*, August 18, 1988; *St. Ignace News*, February 1, 1990. Both newspaper clippings found in RG 98-15, Michigan Department of Transportation, Travel Information and Welcome Centers, box 6, folder "Upper Peninsula," folder 1 of 2, archives of Michigan, Lansing, MI.
27. *St. Ignace News*, July 14, 1988. Newspaper clipping found in RG 98-15, Michigan Department of Transportation, Travel Information and Welcome Centers, box 6, folder "Upper Peninsula," folder 1 of 2, archives of Michigan, Lansing, MI.
28. Petersen, *Inside Mackinac*, 55. Why the two men had not simply taken the ferry

from Mackinac Island to St. Ignace was not explained.

29. *St. Ignace News*, December 12, 1994.
30. David A. Armour, *100 Years at Mackinac: A Centennial History of the Mackinac Island State Park Commission, 1895–1995* (Mackinac Island State Historic Parks, 1995), 82.
31. MISPC minutes August 20, 1960, 636, Keith R. Widder Library, MISPC Office, Mackinaw City, MI.
32. Armour, *100 Years at Mackinac*, 82.
33. Attendance figures 1960 to 1990 courtesy of Mackinac Island State Park Commission.
34. MISCP Minutes, August 4, 1967, 1069; October 18, 1967, 1078.
35. Armour, *100 Years at Mackinac*, 80.
36. Armour, *100 Years at Mackinac*, 80, 95; Daniel M. Spotts, ed., *Travel and Tourism in Michigan: A Statistical Profile*, 2nd ed. (Michigan State University, Travel Tourist and Recreation Center, 1991), 192.
37. Armour, *100 Years at Mackinac*, 98–100. Ticket sales data courtesy of Mackinac Island State Park Commission staff.
38. Victor Hogg, "Mill Creek: An Interpretive Plan Prepared for the Mackinac Island State Park Commission," December 31, 1977, 77–80; attendance figures courtesy of the Mackinac Island State Park staff.
39. Lawrence A. Rubin, *Bridging the Straits: The Story of Mighty Mac* (Wayne State University Press, 1985), 160.
40. Rubin, *Bridging the Straits*, 160–166; *Mackinaw: A Journal for Northern Michigan*, September 12, 1989, 1.
41. Ken Teysen recalls he and a friend, Raymond Desy, invented the idea of the pageant one day over coffee. Teysen makes no reference to the prior Pageant held before World War II. Reprint of interview with Teysen conducted in 2004 by Sandy Planisek, in *Mackinaw News by Sandy Planisek*, May 21, 2023, 12–13.
42. MISPC minutes, October 4, 1962, 801; December 1, 1962, 803; April 24, 1963, 824; Keith R. Widder Library, MISPC Office, Mackinaw City, MI. Larry Rubin, executive directory of the Mackinac Bridge Authority, also seems to have played a role in the rebirth of the pageant. MISPC minutes, April 24, 1963, 824. Larry Rubin had been a collegiate lacrosse player and took a special interest in sponsoring a lacrosse tournament as part of the first event. The Mackinac Bridge Authority remained involved in the pageant until 1970.
43. Philip Deloria, *Playing Indian* (Yale University Press, 1998), 128–129. The hobbyists even had their own publication, *American Indian Hobbyist*, which began publication

in 1954, and its successor, *American Indian Tradition*, both of which frequently encouraged readers to attend pow-wows and similar gathering to learn techniques. A third publication aimed specifically at the traveling audience was *Powwows Trails*, which first appeared in 1964. Deloria, *Playing Indian*, 136, 140.

44. *DFP*, May 23, 1971; MISPC minutes, May 5, 1972, 1314, Keith R. Widder Library, MISPC Office, Mackinaw City, MI.
45. *DFP*, May 21, 1967, 71.
46. *Town Crier*, June 15, 1969, 10.
47. MISPC minutes, June 26, 1970, 1232, Keith R. Widder Library, MISPC Office, Mackinaw City, MI.
48. *DFP*, May 9, 1971, 50.
49. *DFP*, August 24, 1964, 10. Originally, the MBA alternated the direction of the walk each year, but after 1964, the MBA decided to always begin the walk at the bridge's north side (St. Ignace) and terminate the walk at the south end of the bridge (Mackinaw City). Rubin, *Bridging the Straits*, 162.
50. *DFP*, August 14, 1966, 46.
51. Greg Teysen, presentation to the Mackinaw Area Historical Society, January 8, 2024, Mackinaw Area Public Library, Mackinaw City, MI. Mary Lou Peters, *Mackinaw Loves Its Bridge: Memories of the Construction of the Mackinaw Bridge* (Mackinaw Area Historical Society, 2007), 47. It should be added that a Mackinaw City merchant also invented and sold the "Hell No, I didn't walk the Bridge" T-shirt.
52. *Mackinaw: A Journal for Northern Michigan*, April 19, 1983, 9; May 3, 1983, 22. May 17, 1983, 15, reports that the first fudge festival was held in 1980. The same news story adds that while the festival had largely been organized by the Chamber of Commerce, they were asking the town's fudge makers to "step up" and assume greater responsibility for the event. Apparently, they did not and the festival ended. The first Iron Workers festival was held in 1984. See August 7, 1984, 14; August 21, 1984, 4. Because of the relative closeness of Michigan's two peninsulas at the Straits of Mackinac, the area is a major flyway for migratory birds, making it a location also visited by bird-watchers.
53. Department of the Army Corps of Engineers, *Waterborne Commerce of the United States Calendar Year 1960, Part 3 Waterways and Harbors Great Lakes* (n.p., 1962). 159.
54. US Army Corps of Engineers, *Waterborne Commerce of the United States Calendar Year 1989, Part 3 Waterways and Harbors Great Lakes* (Dept. of the Army, Corps of Engineers, Water Resource Support Center, n.d.), 42.
55. P. Donan, "Michigan Central Railroad Tourist Guide, Mackinac Island (1883)," in

Memories of Mackinaw, comp. Judy Ranville and Nancy Campbel, (Little Traverse Printing, 1976), 112.

56. *DFP*, October 30, 1967, 18; Jean R. Beach and Don Steele, *Undefeated: The True Story of How the Family-Owned Shepler's Mackinac Island Ferry Service Survived and Advanced through Three Generations* (AuthorHouse, 2014), 23, 28.
57. Postcard, "Cruiser Pilot II," postcard 798, John Marsh Postcard Collection, Mackinaw Area Public Library, Mackinaw City, MI.
58. Otto W. Lang, "The Reminisces of Otto W. Lang," 1975, Keith R. Widder Library, Mackinaw City, MI, 2, 42–43; Beach and Steele, *Undefeated*, 8; *Mackinac Island News*, August 16, 1941, 3.
59. Beach and Steele, *Undefeated*, 17, 23.
60. Beach and Steele, *Undefeated*, 6, 8, 29–30, 35–36, 37, 39, 42.
61. Department of the Army Corps of Engineers, *Waterborne Commerce of the United States Calendar Year 1958, Part 3 Waterways and Harbors Great Lakes*, 165; *Waterborne Commerce of the United States Calendar Year 1964, Part 3 Waterways and Harbors Great Lakes*, 182.
62. Jaggi Family, "Audie's Restaurant: Family, Food, and Community" (Presentation at the Mackinaw Area Historical Society, Mackinaw City, MI, June 10, 2024); Audie's, https://www.audies.com/about/.
63. Number of hotels and other overnight accommodations in Mackinaw City derived from R. L. Polk, *Cheboygan City and County Directories* (R. L. Polk, 1958, 1961, 1963, 1965, 1967, 1969, 1971, 1973, 1976, 1978, 1980, 1982, 1984, 1986). Although the corporate limits of Mackinaw City fall within two Michigan counties, the Polk Directory for Cheboygan County included listings for business within the village of Mackinaw City found in both Cheboygan and adjoining Emmet County. Contemporary press reports did not always match the number of hotels found in the Polk directory. For example, a 1975 newspaper account claimed Mackinaw City had seventy-six hotels. *DFP*, February 23, 1975, 11. I have chosen to rely on the Polk directory for this analysis.
64. *Mackinaw: A Journal for Northern Michigan*, December 8, 1987, 1, reports the new Best Western Dockside motel was constructed on the Old Breakers Cabin Site. *Mackinaw: A Journal*, July 31, 1990, reports the transformation of the Loveland Motel site into a new Super 8 Motel. *Michigan: A Guide to Motels and Hotels* (Travel Bureau, Michigan Department of Commerce, 1984), 100–104. Motel names selected from those found in the *Cheboygan City Directory, 1986*, R. L. Polk, 1986. Counting the number of motels in Mackinaw City proved challenging. For

example, the *Guide to Motels and Hotels* published by the Travel Bureau in 1984 lists forty-five motels in Mackinaw City while the Polk directory for the same year lists forty-eight. The most likely explanation is that the Polk directory listed every motel in the town including those not currently open for business while the Travel Bureau collected data regarding motels that were open and where the public could book a room. However, it is also possible that the Travel Bureau failed to include a few, likely very small, accommodations.

65. See, for example, *Republican News and St. Ignace Enterprise*, July 7, 1934, 1; March 27, 1937, 1; June 2, 1949, 1; July 14, 1949, 1; August 4, 1949, 1; August 2, 1951, 1; August 12, 1954, 1; July 28, 1955, 1; July 26, 1956, 1; August 15, 1957, 1; June 26, 1958, 1; August 7, 1958, 1; August 6, 1959, 1.
66. *DFP*, August 14, 1966, 46.
67. Unidentified newspaper clipping, dated October 12, 1969, found in folder "St. Ignace," vertical file, Michigan Room, Mackinaw Area Public Library.
68. *DFP*, January 31, 1975, 4.
69. *Town Crier*, June 14, 1980, 4.
70. Sean E. Reavie, *Rebel without a Clue* (Rich Pageant Media, 2022), 89.
71. *DFP*, June 24, 1979, 67.
72. S. Reavie, *Rebel without a Clue*, 90.
73. S. Reavie, *Rebel without a Clue*, 91; Edward K. Reavie, *St. Ignace Car Culture* (Arcadia Press, 2010), 7–8; *DFP*, June 24, 1979, 67. Some sources report that there were 132 cars in the first show. I have used the number provided by Sean Reavie.
74. E. Reavie, *St. Ignace Car Culture*, 7–8, 42; *DFP*, June 24, 1979, 67.
75. S. Reavie, *Rebel without a Clue*, 147–149.
76. S. Reavie, *Rebel without a Clue*, 113–116, 108–109; *St. Ignace News*, February 4, 2010.
77. S. Reavie, *Rebel without a Clue*, 118–121, 124.
78. E. Reavie, *St. Ignace Car Culture*, 86–88, 90; *DFP*, June 29, 1984.
79. S. Reavie, *Rebel without a Clue*, 335, 527.
80. Petersen, *Inside Mackinac*, 76–77. W. Stewart Woodfill to Mackinac Island State Park Commissioners, September 16, 1960, Petersen Woodfill Correspondence, box 1, MISPC Archives, Keith R. Widder Library, MISPC Office, Mackinaw City, MI.
81. Petersen, *Inside Mackinac*, 78.
82. Petersen, *Inside Mackinac*, 78. In discussing the fencing issue, Petersen claims Woodfill never really intended to fence in the entire park but that the leadership of Mackinaw City reacted vehemently to the idea, apparently believing he did mean to do otherwise. Petersen, *Inside Mackinac*, 77–78.

83. Petersen, *Inside Mackinac*, 78–79.
84. Petersen, *Inside Mackinac*, 78–79; W. Stewart Woodfill to Donald Ruffer, September 9, 1960, Petersen Woodfill Correspondence, box 1, MISPC Archives, Keith R. Widder Library, MISPC Office, Mackinaw City, MI.
85. Petersen, *Inside Mackinac*, 77–79. W. Stewart Woodfill to Eugene Petersen, September 1, 1960, Petersen Woodfill Correspondence, box 1, MISPC Archives, Keith R. Widder Library, MISPC Office, Mackinaw City, MI.
86. Armour, *100 Years at Mackinac*, 83.
87. W. Stewart Woodfill to Harry Teysen, September 3, 1966, Petersen Woodfill Correspondence, box 1, MISPC Archives, Keith R. Widder Library, MISPC Office, Mackinaw City, MI.
88. Armour, *100 Years at Mackinac*, 93, 96; MISPC minutes, October 18, 1967, 1079, Keith R. Widder Library, MISPC Office, Mackinaw City, MI; *Town Crier*, August 25, 1968, 1.
89. W. Stewart Woodfill to Walter Murray, March 7, 1962, Petersen Woodfill Correspondence, box 1, MISPC Archives, Keith R. Widder Library, MISPC Office, Mackinaw City, MI.
90. Woodfill may not have been happy about that peace, a process that likely began under his successor as chair, Walter Murray. In a note to Petersen, Woodfill wrote, "I just hope Walter is not taken in by the people of Mackinaw City." Note to Eugene Petersen attached to copy of letter, W. Stewart Woodfill to Walter Murray, March 7, 1962, Petersen Woodfill Correspondence, box 1, MISPC Archives, Keith R. Widder Library, MISPC Office, Mackinaw City, MI.
91. *Town Crier*, August 27, 1983, 14. The mayor of Mackinaw City's official title is president of the village council. Note also that the community of Mackinaw City is, under Michigan law, is not a city but a village.
92. Department of the Army Corps of Engineers, *Waterborne Commerce of the United States Calendar Year 1958*, 165.
93. Beach and Steele, *Undefeated*, 25.
94. *Mackinaw: A Journal of Northern Michigan*, January 8, 1985, 1; April 30, 1985, 2, 4.
95. *Mackinaw: A Journal of Northern Michigan*, February 3, 1987, 2, 4.
96. *Mackinaw: A Journal of Northern Michigan*, March 17, 1987, 2–5, 9–10; January 5, 1988, 5. There was also an argument made that local merchants were using the plan the have someone else pay for various improvements unrelated to the history project that they desired. *Mackinaw: A Journal of Northern Michigan*, March 31, 1987, 4; April 14, 1987, 3.
97. *Mackinaw: A Journal of Northern Michigan*, November 12, 1985, 4; March 5, 1985, 5;

February 3, 1983, 4, 18; March 2, 1983, 5.

98. *Mackinaw: A Journal of Northern Michigan*, February 2, 1988, 2–3.
99. *Republican News and St. Ignace Enterprise*, February 20, 1958, 1.
100. Department of the Army Corps of Engineers, *Waterborne Commerce of the United States Calendar Year 1958*, 166.
101. *Town Crier*, June 30, 1968, 3.
102. Spotts, *Travel and Tourism in Michigan*, 192. Attendance figures courtesy the Mackinac Island State Park Commission staff.
103. S. Reavie, *Rebel without a Clue*, 112.
104. Wes Maurer Jr., quoted in S. Reavie, *Rebel without a Clue*, 522.
105. S. Reavie, *Rebel without a Clue*, 122.
106. E. Reavie as quoted in S. Reavie, *Rebel without a Clue*, 585.
107. S. Reavie, *Rebel without a Clue*, 544, 546–547.
108. S. Reavie, *Rebel without a Clue*, 521–522.
109. *DFP*, October 7, 1976, 19.
110. Beach and Steele, *Undefeated*, 40. Bill Shepler attempted to make Mackinaw City a distinct tourist destination through the development in the 1990s of a project called Mackinaw Crossings. That the local business community failed to support the project in the way Shepler believed was necessary for it to succeed is a story that goes beyond the scope of this book.

EPILOGUE

1. Anna Jameson, *Winter Studies*, in *Historic Mackinac*, Edwin O. Wood (MacMillan Company, 1918), 2:288–295; Margaret Fuller, *Summer on the Lakes*, in Wood, *Historic Mackinac*, 2:363–375.
2. *Detroit Free Press* (hereafter, *DFP*), June 7, 1941, 16; June 22, 1941, 6.
3. For a description of the island's resident native community during the nineteenth century, see Theresa L. Weller, *The Founding Mothers of Mackinac Island: The Agatha Biddle Band of 1870* (Michigan State University Press, 2021), and Melissa Croghan, *Great Women of Mackinac, 1800–1950* (Michigan State University Press, 2023). Croghan's book is not exclusively about the Native community, but she discusses several Anishinaabe women who lived on the Island.
4. John Disturnell, comp., *Island of Mackinac, Giving a Description of All the Objects of Interest and Places of Resort* (Philadelphia, 1875), 95–96; *Mackinac Island News*, July

26, 1941, 8.

5. *Town Crier*, June 19, 1966, 14.
6. *Town Crier*, July 2, 1966, 2. The phrase "imported beadwork" refers to non-Native made items made to look as if they were Native-made, imported from outside of the United States.
7. *Town Crier*, July 17, 1966, 9.
8. *St. Ignace Enterprise*, June 4, 1925, 1; April 21, 1927, 1; August 4, 1927, 1; July 26, 1928, 1; August 29, 1929, 1; St. Ignace Public Library, *St. Ignace* (Arcadia, 2008), 108.
9. *Town Crier*, August 23, 1941, 15.
10. *Town Crier*, June 29, 1991, 14.
11. Paul Stemkowski, "The Fort Gift Shop & Candy Shop," presentation made on September 9, 2024, at the Mackinaw Area Historical Society, Mackinaw City, MI.
12. Judy Ranville, "Chris Schneider and His Park," in *Memories of Mackinaw*, comp. Judy Ranville and Nancy Campbell (Little Traverse Printing, 1976), 85.
13. *DFP*, December 7, 1933, 8.
14. Ranville, "Chirs Schneider," 85.
15. *DFP*, June 28, 1958, 2.
16. *Town Crier*, June 15, 1969, 10. Unidentified newspaper clipping dated June 27, 1977, Mackinaw Area Public Library Notebook, Mackinaw City MI & Vicinity Residents, Michigan History Room, Mackinaw Area Public Library, Mackinaw City, MI.
17. The confrontation between the tourist industry in Mackinaw City and the Native community was one of many occurring in this period throughout the United States. See Philip Deloria, *Playing Indian* (Yale University Press, 1998), 144–146, on this subject.
18. *DFP*, May 23, 1971, 29. The Teysen Indian Museum was not discussed during this controversy.
19. *DFP*, May 23, 1971, 29.
20. Mackinac Island State Park Commission (hereafter, MISPC) minutes, Aril 24, 1963, 824, Keith R. Widder Library, MISPC Office, Mackinaw City, MI.
21. *DFP*, May 9, 1971, 50.
22. MISPC minutes, May 5, 1972, 1314, Keith R. Widder Library, MISPC Office, Mackinaw City, MI.
23. *Town Crier*, June 25, 1971, 8.
24. *Town Crier*, June 24, 1972, 4; MISPC minutes, May 4, 1972, 1314, Keith R. Widder Library, MISPC Office, Mackinaw City, MI. The Order of the Arrow was the organizing group. It is an honor society within the Boy Scout organizational

framework that drew much of its imagery from Native Culture.

25. *Town Crier*, June 15, 1969, 10. The same news story mentioned that a performance by the Chic-a-gami Paleface Indian maidens was rained out. *Chief Wawatam* was the name of the ferry boat that shuttled railroad cars between Mackinaw City and St. Ignace.
26. *Town Crier*, August 11, 1972, 17, 21; undated flyer, "Old Indian Burial Ground, Earl Richardson, proprietor," found in "St. Ignace," vertical file, Michigan History Room, Mackinaw Area Public Library, Mackinaw City, MI.
27. *Town Crier*, August 18, 1972, 20.
28. *DFP*, July 28, 1912.
29. MISPC minutes, June 5, 1901, 37; May 14, 1903, 44; June 17, 1909, 83, Keith R. Widder Library, MISPC Office, Mackinaw City, MI. David A. Armour, *100 Years at Mackinac: A Centennial History of the Mackinac Island State Park Commission, 1895–1995* (Mackinac Island State Historic Parks, 1995), 16–18.
30. MISPC minutes, October 2, 1917, 104, Keith R. Widder Library, MISPC Office, Mackinaw City, MI. Officially, carriage drivers took their guests wherever the guest wished to go. In practice, most guests simply asked the driver to show them the sites, and let the driver pick the route.
31. MISPC Minutes, October 21, 1921, 122; May 23, 1922, 125; Keith R. Widder Library, MISPC Office, Mackinaw City, MI; Armour, *100 Years at Mackinac*, 16.
32. MISPC minutes, May 23, 1922, 125.
33. MISPC minutes, August 19, 1924, 135; October 30, 1924, 137; January 28, 1927, 163; Keith R. Widder Library, MISPC Office, Mackinaw City, MI.
34. MISPC minutes, August 17, 1925, 142, Keith R. Widder Library, MISPC Office, Mackinaw City, MI.
35. MISPC minutes, January 28, 1927, 163; September 7, 1928, 173; Keith R. Widder Library, MISPC Office, Mackinaw City, MI.
36. MISPC minutes, April 27, 1929, 176, Keith R. Widder Library, MISPC Office, Mackinaw City, MI.
37. MISPC minutes, July 5, 1930, 201, Keith R. Widder Library, MISPC Office, Mackinaw City, MI.
38. MISPC minutes, May 5, 1931, 210, Keith R. Widder Library, MISPC Office, Mackinaw City, MI.
39. MISPC minutes, July 10, 1936, 259, Keith R. Widder Library, MISPC Office, Mackinaw City, MI.
40. *Republican News and St. Ignace Enterprise*, September 22, 1938, 8.

41. *Republican News and St. Ignace Enterprise*, May 20, 1948, 2. Some sources date the creation of the Mackinac Island Carriage Tours organization to 1947. See *Republican News and St. Ignace Enterprise*, November 8, 1951, 10. The December 3, 1946, MISPC minutes (Keith R. Widder Library, MISPC Office, Mackinaw City, MI) mention the intention of the operators to form a new organization and suggest the idea was not the commission's but rather the operators.'
42. *Town Crier*, July 13, 1985, 15.
43. MISPC Minutes, June 1, 1950, 379, Keith R. Widder Library, MISPC Office, Mackinaw City, MI.
44. MISPC minutes, September 1, 1951, 389, Keith R. Widder Library, MISPC Office, Mackinaw City, MI.
45. MISPC minutes, December 27, 1951, 391, Keith R. Widder Library, MISPC Office, Mackinaw City, MI.
46. Eugene T. Petersen, *Inside Mackinac* (Inside Mackinac, 1990), 122.
47. MISPC minutes, October 21, 1941, 302, Keith R. Widder Library, MISPC Office, Mackinaw City, MI.
48. Petersen, *Inside Mackinac*, 123.
49. Petersen, *Inside Mackinac*, 124–126.
50. McCabe, *Grand Hotel*, 172.
51. Petersen, *Inside Mackinac*, 125.
52. MISPC minutes, August 4, 1958, 525, Keith R. Widder Library, MISPC Office, Mackinaw City, MI.
53. "Annual Superintendent's Report to the Mackinac Island State Park Commission," December 2, 1960, MISPC, Superintendent Records, box 1, MISPC Archives, Keith R. Widder Library, MISPC Office, Mackinaw City, MI; MISPC minutes, August 30, 1962, 784; October 4, 1962, 796, Keith R. Widder Library, MISPC Office, Mackinaw City, MI.
54. MISPC minutes, November 30, 1962, 725, Keith R. Widder Library, MISPC Office, Mackinaw City, MI.
55. MISPC minutes, April 29, 1967, 1048, Keith R. Widder Library, MISPC Office, Mackinaw City, MI.
56. MISPC minutes, November 14, 1974, 1495, Keith R. Widder Library, MISPC Office, Mackinaw City, MI.
57. Petersen, *Inside Mackinac*, 112–113.
58. *St. Ignace Enterprise*, August 2, 1923, 8.
59. *St. Ignace: The Gate-way City to the Upper Peninsula* (Business Men's Association,

1901). Greg Teysen, presentation to the Mackinaw Area Historical Society, January 8, 2024, Mackinaw Area Public Library, Mackinaw City, MI.

60. *St. Ignace Enterprise*, May 27, 1926, 8.
61. *St. Ignace Enterprise*, August 2, 1923, 8.
62. *St. Ignace Enterprise*, March 16, 1916, 1.
63. *St. Ignace Enterprise*, May 11, 1922, 1.
64. *Republican News and St. Ignace Enterprise*, August 24, 1933, 1.
65. *Republican News and St. Ignace Enterprise*, April 10, 1947, 1.
66. *Republican News and St. Ignace Enterprise*, May 27, 1933, 1.
67. Press release, Mackinaw City Chamber of Commerce, *Mackinaw News*, February 13, 2022, 9.
68. Petersen, *Inside Mackinac*, 75–76. That things would not go well in the future may have been indicated when Woodfill was presented with a proclamation passed by the village council proclaiming him an honorary citizen, a kind gesture but one which might have had greater impact if the person who had prepared the text had not misspelled W. Stewart Woodfill's name as "Stuart" (Petersen, *Inside Mackinac*, 75).
69. Petersen, *Inside Mackinac*, 76–79. W. Stewart Woodfill to Eugene Petersen, September 1, 1960; Woodfill to Donald Ruffer, September 9, 1960; Woodfill to MISPC Commissioners, September 16, 1960; all three letters in Petersen Woodfill Correspondence, box 1, MISPC Archives, Keith R. Widder Library, MISPC Office, Mackinaw City, MI.
70. *Republican News and St. Ignace Enterprise*, June 20, 1940, 1.
71. *Republican News and St. Ignace Enterprise*, July 20, 1940, 1. For information regarding Roger Andrews, see Armour, *100 Years at Mackinac*, 58, 62.
72. *DFP*, July 1, 1979, 90.
73. Some retailers solved this problem by having a second seasonal business in the southern United States, which they operated during the winter months. Although possible, it did involve the expense of maintaining two businesses and two places in which to live.
74. Population statistics for the three communities on the straits came from a variety of publications issued by the State of Michigan and the US Census. Census figures for the years 1894, 1890, 1900, 1910, and 1920 are taken from what is often referred to as the *Michigan Manual*, a publication issued every two years by the Michigan secretary of state primarily to detail those elected to statewide office every two years, but which, over time, became a publication where a great deal of

other information was printed. Consulted were the *Legislative Manual and Official Directory of the State of Michigan for the Year 1885–1886* (Lansing, 1885), 259, 261; *Official Directory and Legislative Manual of the State of Michigan*, 1891–1892 (Lansing, 1989), 649, 660; *Michigan Official Directory and Legislative Manual, 1903–1904* (Robert Smith Printing Co., 1903), 325, 328, 338; *Michigan Official Directory and Legislative Manual, 1913–1914* (n.p., n.d.), 218, 221, 231; *Michigan Official Directory and Legislative Manual, 1923–1924* (n.p., n.d.), 224, 228, 237.

Population Statistics for 1930, 1940, and 1950 were derived from US Department of Commerce, Bureau of the Census, *1950 Census of the Population*, vol. 1, *Number of Inhabitants: Michigan*, 22-15, 22-16, 22-18. Figures for 1960 are from the US Department of Commerce, Bureau of the Census, *1960 Census, Characteristics of the Population*, part 24, *Michigan*, table 7, pages 24-16, 24-18, https://www2.census.gov/library/publications/decennial/1960/population-volume-1/37722966v1p24ch2.pdf; figures for 1970 are from US Department of Commerce, Bureau of the Census, *1970 Census, Population*, vol. 1, *Characteristics of the Population*, part 24, *Michigan*, table 6, https://www.census.gov/library/publications/1973/dec/population-volume-1.html#par_list. Note the US Department of Commerce, Bureau of the Census, *Advance Report, Michigan, January 1971* (PC(v1) -24), table 2, pages 5–6, 10, has a different figure for the population of Mackinac Island in 1960 that does not appear in the census counts for 1960 and 1970. I have ignored this variant. The count for 1980 is from US Department of Commerce, Bureau of the Census, *Local Population Estimates, East North Central Series*, part 26, no. 84 ENC-SC, issued June 1986, table 1, pages 62, 64, 70. Figures for 1990 are from "Mackinaw City Population History, 1990–2023," Biggest US Cities, https://www.biggestuscities.com/; "Mackinac Island Population History 1990–2023," Biggest US Cities, https://www.biggestuscities.com/city/mackinac-island-michigan; and "St. Ignace Population History, 1990–2023," Biggest US Cities, https://www.biggestuscities.com/city/st-ignace-michigan. For the purposes of this book, the small populations of nearby townships are not included.

75. Population statistics for the three communities on the straits came from a variety of publications issued by the State of Michigan and the US Census. Census figures for the years 1894, 1890, 1900, 1910, and 1920 are taken from what is often referred to as the *Michigan Manual*, a publication issued every two years by the Michigan secretary of state primarily to detail those elected to statewide office every two years, but which, over time, became a publication where a great deal of

other information was printed. Consulted were the *Legislative Manual and Official Directory of the State of Michigan for the Year 1885–1886*, 259, 261; *Official Directory and Legislative Manual of the State of Michigan, 1891–1892*, 649, 660; *Michigan Official Directory and Legislative Manual, 1903–1904*, 325, 328, 338; *Michigan Official Directory and Legislative Manual, 1913–1914*, 218, 221, 231; *Michigan Official Directory and Legislative Manual, 1923–1924*, 224, 228, 237.

Population Statistics for 1930, 1940, and 1950 were derived from US Department of Commerce, Bureau of the Census, *1950 Census of the Population*, vol. 1, *Number of Inhabitants: Michigan*, 22-15, 22-16, 22-18. Figures for 1960 are from the US Department of Commerce, Bureau of the Census, *1960 Census, Characteristics of the Population*, part 24, *Michigan*, table 7, pages 24-16, 24-18, https://www2.census.gov/library/publications/decennial/1960/population-volume-1/37722966v1p24ch2.pdf.

76. *DFP*, September 3, 1962, 222; April 2, 1967, 3–4. MRA was extremely controversial on the island, and concluding that some of its island supporters supported it for personal economic reasons must be drawn indirectly from the available evidence.
77. McCabe, *Grand Hotel*, 142.
78. Harriet Martineau, *Society in America*, as quoted in Wood, *Historic Mackinac*, 2:268–269.

Bibliography

ARCHIVES AND COLLECTIONS

Mackinac Island State Park Commission Archives, Keith R. Widder Library, Mackinaw City, MI.

Mackinac Island State Park Commission minutes, Keith R. Widder Library, Mackinaw City, MI, and the Michigan History Room, Mackinaw Area Public Library, Mackinaw City, MI.

Petersen, Eugene, and W. Stewart Woodfill Correspondence, Keith R. Widder Library, Mackinaw City, MI. Includes copies of letters sent to other individuals usually written by Woodfill, dated as early as 1946.

Travel Information and Welcome Centers, RG 98-15. Bureau of Michigan History, Archives of the State of Michigan, Lansing, MI.

White, Peter, Papers, Bentley Historical Library, University of Michigan.

Williams, G. Mennen, Papers, Bentley Historical Library, University of Michigan.

PUBLICATIONS

Adie, Madeline Okerman. *Mackinaw City*. Arcadia, 2007.

Adie, Madeline Okerman. *The Straits of Mackinac*. Arcadia, 2012.

Agassiz, Louis. *Lake Superior: Its Physical Character, Vegetation, and Animals Compared with Those of Other and Similar Regions*. With a narrative of the tour by J. Elliot Cabot. Boston, 1850.

Alsberg, Henry G., ed. *The American Guide: A Source Book and Complete Travel Guide for the United States*. Hastings House, 1949.

Anderson, David D. "John Disturnell Introduces the Great Lakes to America." *Inland Seas* 18 (Summer 1962): 96–106.

Andrews, Roger. "Frontier Restaurant: A Teysen Tradition." Originally printed in the Mackinac Island News, July 26, 1941. In Ranville and Campbell, *Memories of Mackinaw*, 301–302.

Andrews, Roger. *Old Fort Mackinac on the Hill of History*. Herald-Leader Press, 1938.

"An Interview with R.D. Musser II." Grand Hotel. https://www.grandhotel.com/press-media/interview-r-d-musser-ii/.

"Annual Re-enactment Pontiac Massacre of 1763, June 30, 1934." Flyer in Ranville and Campbell, *Memories of Mackinaw*, 91.

Appletons' Hand-Book of American Travel, Western Tour. New York, 1871.

Armour, David A. *100 Years at Mackinac: A Centennial History of the Mackinac Island State Park Commission, 1895–1995*. Mackinac Island State Historic Parks, 1995.

Aron, Cindy S. *Working at Play: A History of Vacations in the United States*. Oxford University Press, 1999.

Austin, H. W. "Bunny." *Frank Buchman as I Knew Him*. Grosvenor Books, 1975.

Bailey, Robert Matthew. "Reminisces of Robert Matthew Bailey and Philip Keast Truscott." Interview by Thomas Pfeiffelmann, February and March 1973. Keith R. Widder Library, Mackinaw City, MI.

Barbour, Levi L. "Peter White as Man and as Citizen." *Michigan Alumnus* 15 (1909): 407–412.

Bard, Camden. "North: A History of Natural and Imagined Landscapes in the Great Lakes Region." MA thesis, Central Michigan University, 2014.

Barnett, Le Roy. *Shipping Literature of the Great Lakes*. Michigan State University Press, 1992.

Barrett, William Henry. "The History of Transportation across the Straits of Mackinac." MA thesis, University of Hawaii, 1963.

Beach, Jean R., and Don Steele. *Undefeated: The True Story of How the Family-Owned Shepler's Mackinac Island Ferry Service Survived and Advanced through Three Generations*. AuthorHouse, 2014.

Bederman, Gail. *Manliness and Civilization: A Cultural History of Gender and Race in the United States, 1880–1917*. University of Chicago Press, 1995.

Belasco, Warren James. *American on the Roads: From Autocamp to Motel, 1910–1945*. MIT Press, 1979.

Benjamin, Robert E. *Mackinac Island: Three Hundred Years of History*. Guelff, 1952.

Breighner, Kathryn Washburn. "A Tale of Two Cities." In *Mackinac: The Gathering Place*, edited by Russell McKee, 81–90. Michigan Natural Resources Magazine, 1981.

"Bridging the Straits Museum and Welcome Center, St. Ignace." Site plan for Mackinac Bridge, June 1988, in Bureau of Michigan History, State Archives. RG 98-15.

"Bridging the Straits: A Summary Report Concerning the Feasibility of a Proposed Upper Peninsula Welcome Center/Straits of Mackinac Transportation Museum." Presented by the Mackinac Island State Park Commission, August 15, 1988, in Bureau of MI History, State Archives. RG 9815.

Brisson, Steven C. *Downtown Mackinac Island: An Album of Historic Photos*. Mackinac Island State Park Commission, 2015.

Brisson, Steven. "Mackinac Island State Park, Mackinac Island, Michigan." In *Warren H. Manning, Landscape Architect and Environmental Planner*, edited by Robin Karson, Jane Roy Brown, and Sarah Allaback, 183–186. University of Georgia, 2017.

Brisson, Steven. "Michilimackinac State Park, Mackinaw City, Michigan." In *Warren H. Manning, Landscape Architect and Environmental Planner*, edited by Robin Karson, Jane Roy Brown, and Sarah Allaback, 187–190. University of Georgia, 2017.

Brisson, Steven C. *Wish You Were Here: An Album of Vintage Mackinac Postcards*. Mackinac State Historic Parks, 2002.

Brown, Leona. "Mackinaw City in the Old Days, about 1908." In Ranville and Campbell, *Memories of Mackinaw*, 226–230.

Bryant, William Cullen. *Letters of a Traveller* (1846). In Wood, *Historic Mackinac*, 377–402.

Bryant, William Cullen, ed. *Picturesque America or the Land We Live In*. 2 vols. New York, 1872–1874.

Burd, Camden. "Imagining a Pure Michigan Landscape: Advertises, Tourists, and the Making of Michigan's Northern Vacationlands." *Michigan Historical Review* 42, no. 2 (Fall 2016): 31–51.

Burd, Camden. "In the Land of Hiawatha." In *Pieces of the Heartland: Representing*

Midwestern Places, edited by Andy Oder. Hastings College Press, 2018.

Burns, Benjamin. "A Political History of the Mackinac Bridge." MA thesis, Michigan State University, 1968.

Burton, Clarence M., and M. Agnes Burton. *History of Wayne County and the City of Detroit, Michigan*. 3 vols. S. J. Clarke, 1930.

Byron, M. Christine, and Thomas R. Wilson. *Vintage Views Along Scenic M-22 Including Sleeping Bear Dunes*. Vintage Views Press, 2015.

Byron, M. Christine, and Thomas R. Wilson. *Vintage Views along the West Michigan Pike: From Sand Trails to US-31*. Arbutus Press, 2011.

Byron, M. Christine, and Thomas R. Wilson. *Vintage Views of the Mackinac Straits Region*. Arbutus Press, 2007.

Campbell, Steven R. "The Development of Mackinaw City." In Ranville and Campbell, *Memories of Mackinaw*, 51–53.

Cantor, George. *Explore Michigan: Mackinac*. University of Michigan, 2005.

Cantor, George. *Old Roads of the Midwest*. University of Michigan Press, 1997.

Census of the State of Michigan, 1874. Lansing, 1875.

Census of the State of Michigan, 1884. 2 vols. Lansing, 1886.

Centennial History of Cheboygan County. Northern Tribune Print, 1876. Reprint, Beard, 1976.

City of St Ignace and Its Surroundings, Its Attractions for Tourist, Health Seeker, Historian, Traveler and Visitor. Business Men's Association, 1901.

Clayton, James L. "The American Fur Company: The Final Years." PhD diss., Cornell University, 1964.

Coffman, Edna M. *Mackinaw City Settlers and the Savage Straits*. Voyager Press, 1976.

Congressional Record Containing the Proceedings and Debates of the 43rd Congress, Special Session of the Senate of the United States, March 4 to 26, 1873. Vol. 1. US Congress, 1874.

Coverdale & Colpitts. "Report on Traffic and Revenues Proposed Mackinac Straits Bridge." New York, January 22, 1952.

Croghan, Melissa. *Great Women of Mackinac, 1800–1950*. Michigan State University Press, 2023.

Curtiss, Daniel S. *Western Portraiture: An Emigrants' Guide*. New York, 1852.

Davey, Joseph F. "Memories of Mackinac." Unpublished manuscript, 1982. Keith R. Widder Library, Mackinaw City, MI.

Davis, Marion Morse. "Stories of Saint Helena Island." *Michigan History Magazine* 10 (1926), 411–446.

Deloria, Philip. *Playing Indian*. Yale University Press, 1998.

Dempsey, Dave. *Ruin and Recovery: Michigan's Rise as a Conservation Leader*. University of Michigan Press, 2001.

Desy, Mary. "The Desy Family." In Ranville and Campbell, *Memories of Mackinaw*, 236–238.

Detroit and Cleveland Steam Navigation Company. *A Hero of Manilla*. Detroit, 1899.

Detroit and Cleveland Steam Navigation Company. *Hunting and Fishing Resorts of the Great Lakes via Picturesque Mackinac*. Buffalo, 1887.

Detroit and Cleveland Steam Navigation Company. *The Mystery of the Monogram: Story of a Summer Tour on the Great Lakes*. John Bornman & Sons, 1904.

Dilworth, Leah. *Imagining Indians in the Southwest: Persistent Visions of a Primitive Past*. Smithsonian Institution Press, 1996.

Disturnell, John, comp. *Island of Mackinac, Giving a Description of All the Objects of Interest and Places of Resort*. Philadelphia, 1875.

Dixie Highway 10, no. 6 (July 1924).

Doerr, Mary Jane. *Bay View, an American Idea*. Priscilla, 2010.

Donan, P. "Michigan Central Tourist Guide, Mackinac Island, 1883." In Ranville and Campbell, *Memories of Mackinaw*, 112.

Doyle, W[ilfird]. F. Interview by Tom Downs, April 13, 1984. Keith R. Widder Library, Mackinaw City, MI.

Drake, Daniel. *The Northern Lakes: A Summer Resort for Invalids of the South*. Louisville, KY, 1842. Reprint, Torch Press, 1954.

Duffina, Mary G. *The Voice of the Turtle: A Pictorial Narrative of Early Mackinac Island Formerly Michilimackinac*. Mary G. Duffina, 1991.

Dunbar, Willis F. *All Aboard! A History of Railroads in Michigan*. William Eerdmans, 1969.

Dunnigan, Brian Leigh. *A Picturesque Situation: Mackinac before Photography, 1616–1860*. Wayne State University Press, 2008.

Dunnigan, James P. "Around My World in Eighty Years." Unpublished memoir, December 1990. Keith R. Widder Library, Mackinaw City, MI.

Eby, Clarence C. *Mackinac County of the Straits Country: An Official Tourist Guide into the Upper Peninsula of Michigan*. Clarence C. Eby, 1926.

Emery, Benjamin Franklin. *Sketches of Old Michigan Forts* Vol. 3, *Fort Michilimackinac, 1715–1780, the Scene of the Pontiac Massacre, June 4, 1763*. The Old Forts and Memorials Historic Association, 1931.

Engelmann, Larry. "O Whisky: A History of Prohibition in Michigan." PhD diss., University of Michigan, 1971.

Fairborn, John Baldwin, II. "John Fairborn and the Martel Furnace: The Changing Character of the Entrepreneur in Industry, 1865–1887." BA honors thesis, Middlebury College, 1984. Copy located in the St. Ignace Public Library.

Fornes, Mike. *Mackinac Bridge: A 50 Year Chronicle, 1957–2007*. Cheboygan Tribune, 2007.

Frank, Robin Jaffee. *Coney Island: Visions of an American Dreamland, 1861–2008*. Yale University Press, 2015.

Friday, Matthew J. *Among the Sturdy Pioneers: The Birth of the Cheboygan Area as a Lumbering Community, 1778–1935*. Trafford, 2006.

Fuller, George, ed. *Historic Michigan: Land of the Great Lakes*. Vol. 3, *A History of the Upper Peninsula of Michigan*. National Historical Association, n.d.

Fuller, Margaret. *Summer on the Lakes* (1844). In Wood, *Historic Mackinac*, 361–376.

Grand Rapids & Indiana Railroad. *A Guide to the Health, Pleasure and Fishing Resorts of Northern Michigan*. Grand Rapids, 1879.

Harding, Lewis A., ed. *History of Decatur County, Indiana: Its People, Industries, and Institutions*. B. F. Bowen, 1915.

Haskell, H. Morgan. *W. Stewart Woodfill: Master of Mackinac's Grand Hotel*. H. Morgan Haskell, 2001.

Hazard, Willis P. *The American Guide-Book: Being a Handbook for Tourists and Travellers through Every Part of the United States*. D. Appleton, 1846.

Heasley, Lynne. *The Accidental Reef and Other Ecological Odysseys in the Great Lakes*. Michigan State University Press, 2021.

Henderson, Jennifer. *Settler Feminism and Race Making in Canada*. University of Toronto Press, 2003.

Hoganson, Kristin L. *Consumers' Imperium: The Global Production of American Domesticity, 1865–1920*. University of North Carolina Press, 2007.

Hogg, Victor. "Mill Creek: An Interpretive Plan." Circulated among staff, Mackinac Island State Park Commission. December 31, 1977.

Jackson, Harvey H., III. *Redneck Riveria: An Insider's History of the Florida-Alabama Coast*. University of Georgia Press, 2012.

Jackson, Mark. *Allergy: The History of a Modern Malady*. Reaktion Books, 2006.

Jaggi Family. "Audie's Restaurant: Family, Food, and Community." Presentation at the Mackinaw Area Historical Society, Mackinaw City, MI, June 10, 2024.

Jakle, John A. *The Tourist: Travel in Twentieth-Century North America*. University of Nebraska Press, 1985.

Jameson, Anna. *Winter Studies and Summer Rambles in Canada* (1838). In Wood, *Historic*

Mackinac, 270–299.

Kachadurian, Thomas. *Views of Mackinac Island*. Sleeping Bear Press, 2000.

Kasson, John F. *Amusing the Million: Coney Island at the Turn of the Century*. Hill & Wang, 1978.

Kelton, Dwight H. *Annals of Fort Mackinac*. Chicago, 1882.

Korn, Claire V. *Yesterday through Tomorrow: Michigan State Parks*. Michigan State University Press, 1989.

Kulsea, Bill, and Tom Shawver. *Making Michigan Move: A History of Michigan Highways and the Michigan Department of Transportation*. Michigan Department of Transportation, 1992.

Lang, Otto W. "The Reminisces of Otto W. Lang," 1975. Keith R. Widder Library, Mackinaw City, MI.

Lanman, Charles. *A Summer in the Wilderness*. (1847). In Wood, *Historic Mackinac*,60–65.

Mackinac Island Civic Association. *Mackinac Island*. Mackinac Island Civic Association, 1926.

Mackinac Island State Park Commission. "1933 and 1934 report of the Mackinac Island State Park Commission." Report submitted December 1, 1934.

Mackinac Island State Park Commission. "1937–1938 Report of the Mackinac Island State Park Commission of Michigan, 1938."

Magnaghi, Russell M. *Prohibition in the Upper Peninsula*. American Palate, 2017.

Marquis, Albert Nelson, ed. *The Book of Detroiters: A Biographical Dictionary of Leading Living Men of the City of Detroit*. A. N. Marquis, 1908.

Marryat, Frederick. *Diary in America*. Edited by Jules Zanger. Nicholas Vane, 1960.

Martineau, Harriet. *Society in America* (1837). In Wood, *Historic Mackinac*, 255–269.

May, George. *Historic Guidebook: Mackinac Island*. Mackinac Island State Park Commission, 1962.

May, Marvin. "Reminisces of Marvin May." 1999. Keith R. Widder Library, Mackinac Island State Park Commission, Mackinaw City, MI.

McCabe, John. *Grand Hotel Mackinac Island*. Lake Superior State College, 1987.

McKee, Russell, ed. *Mackinac: The Gathering Place*. Michigan Natural Resources Magazine, 1981.

McKinnon, A. Donald. *A Prairie Farm Boy's Odyssey*. Self-published, ca. 2003. Located in St. Ignace Public Library.

McMahon, Daniel. "The Role of Security Concerns in the of Mackinac National Park." *Michigan Historical Review* 50, no. 2 (Fall 2024): 87–109.

Meints, Graydon M. *The Fishing Line: A History of the Grand Rapids & Indiana Railroad*.

Michigan State University Press, 2018.

Michigan: A Guide to Motels and Hotels. Travel Bureau, Michigan Department of Commerce, 1984.

Michigan Central Railroad Company. *Report of the President and Directors of the Michigan Central Railroad Company to the Stockholders for the Year Ending December 31, 1881*. 1882.

Michigan Department of Commerce, Travel Bureau. *Michigan: A Guide to Hotels and Motels*. Ca. 1984.

Michigan Department of Commerce, Travel Bureau. *Say Yes to Michigan! A Guide to Lodging in Michigan*. Ca. 1982.

Michigan Farming Lands! Broadside issued by the Detroit, Mackinac & Marquette Railroad Co., W. O. Strong, land commissioner, January 30, 1882, Clarke Historical Library, Central Michigan University.

Michigan Historical Commission. *Lewis Cass Day on Mackinac Island*. Bulletin no. 7. Wynkoop Hallenbeck Crawford, 1916.

Michigan Historical Commission. *Nicolet Day on Mackinac Island*. Bulletin no. 6. Wynkoop Hallenbeck Crawford, 1916.

Moore, John, H. "A Reporter at Large: A Little Journey to Mackinaw." In Ranville and Campbell, *Memories of Mackinaw*, 78–80.

Musser, R. Daniel, Jr. "Summer 2012: An Oral History." Conversation with Phil Porter, Mackinac State Historic Parks, Mackinac Island State Park Commission, 2021. Keith R. Widder Library, MISPC Office, Mackinaw City, MI.

Noer, Thomas. J. *Soapy: A Biography of G. Mennen Williams*. University of Michigan Press, 2005.

Nold, Carl R. "The Michigan Governor's Summer Residence." *Mackinac History* 5, leaflet no 6 (2024).

"Official Program, Opening Old Fort Michilimackinac Stockade, Mackinaw City, July 1st, 1933." In Ranville and Campbell, *Memories of Mackinaw*, 93.

O'Malley, John William. "The Story of the West Baden Springs Hotel." MA thesis, Loyola University of Chicago, 1957.

Patrick [Riggs], Marie. "Memories of Yesteryear." In Ranville and Campbell, *Memories of Mackinaw*, 231–233.

Peters, Kenneth. "The Good-Roads Movement and the Michigan State Highway Department, 1905–1917." PhD diss., University of Michigan, 1972.

Peters, Mary Lou. *Mackinaw Loves Its Bridge: Memories of the Construction of the Mackinaw Bridge*. Mackinaw Area Historical Society, 2007.

Petersen, Eugene T. *Inside Mackinac*. Inside Mackinac, 1990.

Petersen, Eugene T. *Mackinac's Grand Hotel: The Early Years*. Historical Society of Michigan, 1988.

Petersen, Eugene T. *Mackinac Island: Its History in Pictures*. Mackinac Island State Park Commission, 1973.

Petersen, Eugene T. *Michilimackinac: Its History and Restoration*. Rev. ed. Mackinac Island State Park Commission, 1968.

Pfeiffelmann, Herbert Eugene. "The Reminiscences of Herbert Eugene Pfeiffelmann." [1978?]. Mackinac Island State Park Commission Library.

Phillips, Katrina M. *Staging Indigeneity: Salvage Tourism and the Performance of Native American History*. University of North Carolina Press, 2021.

Piljac, Pamela A., and Thomas M. Piljac. *Mackinac Island: Historic Frontier, Vacation Resort, Timeless Wonderland*. Bryce-Waterton Publications, 1988.

Porter, Phil. *A Desirable Station: Soldier Life at Fort Mackinac, 1867–1895*. Mackinac State Historic Parks, 2003.

Porter, Phil. *Fudge: Mackinac's Sweet Souvenir*. Mackinac State Historic Parks, 2001.

Porter, Phil. *Mackinac: An Island Famous in These Regions*. Mackinac State Historic Parks, 1998.

Porter Phil. *View from the Veranda: The History and Architecture of the Summer Cottages of Mackinac Island*. Mackinac Island State Park Commission, 1981.

Porter, Phil. "Where Horse Is King: Mackinac Island's Automobile Ban." *Mackinac History* 5, no. 5 (2023).

Porter, William Clan. "Reminisces." 2000. Mackinac Island State Park Commission Library.

Prather, Donald Fry. *There Will Always Be a Mackinac Race*. D. F. Keller, 1925.

Puttkammer, Ernest Wilfred. "Reminisces." 1973. Mackinac Island State Park Commission Library.

Ramin, Gerald A. "Tourism as a Major Resource of Mackinac Island." Independent study, MS degree, Mankato State University, 1963.

Ranville, Judy. "Chris Schneider and His Park." In Ranville and Campbell, *Memories of Mackinaw*, 84–87.

Ranville, Judy. "Ralph Preston's Railroad Memories of 1924." In Ranville and Campbell, *Memories of Mackinaw*, 110–111.

Ranville, Judy, and Nancy Campbell, comps. *Memories of Mackinaw*. Little Traverse Printing, 1976.

Reavie, Edward K. *St. Ignace Car Culture*. Arcadia Publishing, 2010.

Reavie, Sean E. *Rebel without a Clue*. Rich Pageant Media, 2022.

Richards, Thomas Addison. *Appletons' Illustrated Handbook of American Travel*. D. Appleton, 1857.

Ritchie, Robert C. *The Lure of the Beach: A Global History*. University of California Press, 2021.

Rogers, Frank F. *History of the Michigan State Highway Department, 1905–1933*. Franklin DeKleine, 1933.

Rubin, Lawrence A. *Bridging the Straits: The Story of Mighty Mac*. Wayne State University Press, 1985.

Sack, Daniel. *Moral Re-Armament: The Reinventions of an American Religious Movement*. Palgrave Macmillan, 2009.

Sandoval-Strausz, A. K. *Hotel: An American History*. L.E.G.O. SpA, 2007.

Schenck, Noella La Chance. *Winter Park's Old Alabama Hotel*. Anna Publishing, 1982.

Schuster, David G. "Neurasthenia and a Modernizing America." *JAMA* 2907, no. 17 (2003): 2327–2328.

Sears, John F. *Sacred Places: American Tourist Attractions in the Nineteenth Century*. Oxford University Press, 1989.

Shaffer, Marguerite S. *See America First: Tourism and National Identity, 1880–1940*. Smithsonian Institution Press, 2001.

Shapiro, Aaron. *The Lure of the North Woods: Cultivating Tourism in the Upper Midwest*. University of Minnesota Press, 2013.

Shapiro, Aaron. "Up North on Vacation: Tourism and Resorts in Wisconsin's North Woods, 1900–1945." *Wisconsin Magazine of History* 89, no. 4 (Summer 2006): 2–13.

Shepler, Chris. "History of the Shepler Ferries." Presentation to the Mackinaw Area Historical Society, May 13, 2024.

Shine, Agnes. "Agnes Shine Reminisce." As told to Thomas Pfeiffelmann, June 4, 1973. Mackinac Island State Park Commission Library.

Smith, Emerson R., comp. *Before the Bridge: A History and a Directory of St. Ignace and Nearby Localities*. Kiwanis Club of St. Ignace, 1957.

Smith, Marne. "My Life Growing Up in a Small Town [Mackinaw City]." Unpublished memoir, 1980. Michigan History Room, Mackinaw Area Public Library, Mackinaw City, MI.

Spotts, Daniel M., ed. *Travel and Tourism in Michigan: A Statistical Profile*. Michigan State University, Travel, Tourism and Recreation Resource Center, 1986.

Spotts, Daniel M, ed. *Travel and Tourism in Michigan: A Statistical Profile*. 2nd ed. Michigan State University, Travel Tourist and Recreation Center, 1991.

Stemkowski, Paul. "The Fort Gift Shop & Candy Shop." Presentation to the Mackinaw Area Historical Society, Mackinaw City, MI, September 9, 2024.

St. Ignace: The Gate-way City to the Upper Peninsula. Business Men's Association, 1901.

St. Ignace Public Library. *St. Ignace*. Arcadia, 2008.

Strickland. *Old Mackinaw*. In Wood, *Historic Mackinac*, 573–576.

Stites, Susan, and Lea Ann Sterling. *Historic Cottages of Mackinac Island*. Arbutus Press, 2001.

Teysen, Greg. "History of the Teysen Family Business in Mackinaw City." Presentation to the Mackinaw Area Historical Society, Mackinaw City, MI, January 8, 2024.

Thomas, Clara. "Anna Jameson: Art Historian and Critic." *Woman's Art Journal* 1, no. 1 (1980): 20–22.

Travel Michigan, Michigan Economic Development Corporation. *The Michigan Travel and Tourism Handbook, 1999*. Updated 4/99. No other publication data.

"The Traverse Region." In Ranville and Campbell, *Memories of Mackinaw*, 9–13.

Truscott, Philip Keast. "Reminisces of Robert Matthew Bailey and Philip Keast Truscott." Interview by Thomas Pfeiffelmann, February and March 1973. Keith R. Widder Library, Mackinaw City, MI.

Varney, Peter. *Motor Camping*. Leisure League of America, 1935.

Wallin, Lena E. Interview by Judy La Way Ranville. In Ranville and Campbell, *Memories of Mackinaw*, 266–267.

Watson, Andrew. *Making Muskoka Tourism, Rural Identity, and Sustainability*. University of British Columbia Press, 2022.

Wa-Wa-Tam Association. "Wawatam Memories." In Ranville and Campbell, *Memories of Mackinaw*, 66–69.

Weller, Theresa L. *The Founding Mothers of Mackinac Island: The Agatha Biddle Band of 1870*. Michigan State University Press, 2021.

Widder, Keith R. "Mackinac National Park 1875–1975." *Reports in Mackinac History and Archaeology*, no. 4. Mackinac State Park Commission, 1975.

Wilkinson, Walter B. *Profits in Tourist Camps*. Travel Research, 1940.

Williams, Ralph D. *The Honorable Peter White: A Biographical Sketch of the Lake Superior Iron Country*. Penton, 1907.

Williams, W. *Appletons' New and Complete Guidebook for Travellers*. D. Appleton, 1851.

Williams, W. *Appletons' Southern and Western Travellers' Guide*. D. Appleton, 1850.

Wilson, Jack S. "The Travel Market for Michigan." Michigan Department of Commerce, Travel Bureau, 1981. Michigan Documents Collection, Library of Michigan, Lansing, MI.

Wood, Edwin O., ed. *Historic Mackinac: The Historical, Picturesque, and Legendary Features of Mackinac Country*. 2 vols. MacMillan, 1918.

Woodfill, W. Stewart. *The Grand Hotel: The Story of an Institution*. Newcomen Society in North America, 1969.

Index